Digital Curricula in School Mathematics

A volume in
Research in Mathematics Education
Denisse R. Thompson, Mary Ann Huntley, and Christine Suurtamm, *Series Editors*

Digital Curricula in School Mathematics

edited by

Meg Bates
The University of Chicago

Zalman Usiskin
The University of Chicago

INFORMATION AGE PUBLISHING, INC.
Charlotte, NC • www.infoagepub.com

Library of Congress Cataloging-in-Publication Data

A CIP record for this book is available from the Library of Congress
http://www.loc.gov

ISBN: 978-1-68123-411-3 (Paperback)
 978-1-68123-412-0 (Hardcover)
 978-1-68123-413-7 (ebook)

Copyright © 2016 Information Age Publishing Inc.

All rights reserved. No part of this publication may be reproduced, stored in a retrieval system, or transmitted, in any form or by any means, electronic, mechanical, photocopying, microfilming, recording or otherwise, without written permission from the publisher.

Printed in the United States of America

CONTENTS

Preface ix
Acknowledgments xi

1 Welcoming Remarks 1

PART I
CREATING DIGITAL CURRICULUM

2 Designing Curriculum for Digital Middle Grades Mathematics:
Personalized Learning Ecologies 7
Jere Confrey

3 Developing and Implementing "Smart" Mathematics
Textbooks in Korea: Issues and Challenges 35
Hee-chan Lew

4 Technology-Enhanced Teaching/Learning at a New Level
With Dynamic Mathematics as Implemented in the New Cabri 53
Jean-Marie Laborde

5 The Re-Sourcing Movement in Mathematics Teaching: Some
European Initiatives 75
Kenneth Ruthven

6 Inquiry Curriculum and E-Textbooks: Technological Changes
That Challenge the Representation of Mathematics Pedagogy 87
Michal Yerushalmy

PART II
IMPLEMENTING DIGITAL CURRICULUM

7 Connections and Distinctions Among Today's Digital Innovations and Yesterday's Innovative Curricula 109
   Valerie L. Mills

8 Technology to Support Mathematics Instruction: Examples From the Real World 123
   Loretta J. Asay

9 We Thought We Knew It All 133
   Josephus Johnson

10 Deeply Digital Curriculum for Deeply Digital Students 139
   Brian Lemmen

PART III
RESEARCHING DIGITAL CURRICULUM

11 Analysis of Eight Digital Curriculum Programs 161
   Jeffrey Choppin

12 A Design Experiment of a Deeply Digital Instructional Unit and Its Impact in High School Classrooms 177
   Alden J. Edson

13 Keeping an Eye on the Teacher in the Digital Curriculum Race..... 195
   Janine Remillard

14 New Starting Points for Number Sense Using TouchCounts 205
   Nathalie Sinclair

PART IV
BROADER CONSIDERATIONS ABOUT DIGITAL CURRICULUM

15 Digitally Enhanced Learning 225
   Philip Daro

16 Mathematics Standards and Curricula Under the Influence of Digital Affordances: Different Notions, Meanings, and Roles in Different Parts of the World 239
   Mogens Niss

17 Mathematics Curriculum, Assessment, and Teaching for Living in the Digital World: Computational Tools in High Stakes Assessment 251
   Kaye Stacey

18 Mathematics Education is at a Major Turning Point 271
   David Moursund

19 Deeply Digital STEM Learning 285
   Chad Dorsey

20 Closing Remarks 297
   Zalman Usiskin

   Appendix A: Conference Program 303
   Appendix B: Speaker Biographies 309
   Appendix C: Conference Personnel 317

PREFACE

This volume contains the proceedings of the Third International Curriculum Conference sponsored by the Center for the Study of Mathematics Curriculum (CSMC), held November 7–9, 2014 on the campus of the University of Chicago.

The CSMC is one of the National Science Foundation Centers for Learning and Teaching (Award No. ESI-0333879). As noted on the CSMC website (http://mathcurriculumcenter.org), the CSMC serves the K–12 educational community by focusing scholarly inquiry and professional development around issues of mathematics curriculum. Major areas of work include understanding the influence and potential of mathematics curriculum materials, enabling teacher learning through curriculum material investigation and implementation, and building capacity for developing, implementing, and studying the impact of mathematics curriculum materials. The work of the CSMC is not driven by a particular philosophy or ideology. As researchers, CSMC staff try to maintain a healthy skepticism throughout all phases of the center's activities considering both international and multiple U.S. perspectives with an ultimate goal to produce research-based knowledge and products that will enlighten and serve the range of users of mathematics curriculum materials.

The venue, the University of Chicago, was also the location of the first and second CSMC International Conferences held in 2005 and 2008. The first conference was directed at learning about recent trends in four carefully chosen countries, as its title, *Mathematics Curriculum in Pacific Rim Countries—China, Japan, Korea, and Singapore*, suggests.

The success of the first conference led us to structure the second conference in a way analogous to the first. Instead of four countries, we chose four topics: early algebra, computer algebra systems (CAS), 3-dimensional geometry, and linking algebra and geometry. This conference was entitled *Future Curricular Trends in School Algebra and Geometry*. Proceedings of both conferences were also published by Information Age Publishing (Usiskin, Andersen & Zotto, 2010; Usiskin & Willmore, 2008).

This third conference tackled digital curriculum creation, implementation, and research. It began with eight presentations of various uses of digital technology in mathematics and ended with two looks at the future. In this volume are chapters from all the speakers at the conference. It thus makes available the work of the conference to curriculum developers and researchers from around the world who are tackling this issue of digital mathematics curriculum but who were unable to attend the conference. The wider community engaged in this work is able to learn from and continue conversations with other interested individuals.

REFERENCES

Usiskin, Z., Andersen, K., & Zotto, N. (Eds.). (2010). *Future curricular trends in school algebra and geometry: Proceedings of a conference.* Charlotte, NC: Information Age.

Usiskin, Z., & Willmore, E. (Eds.) (2008). *Mathematics curriculum in Pacific Rim countries—China, Japan, Korea, and Singapore.* Charlotte, NC: Information Age.

ACKNOWLEDGMENTS

A conference of this type requires the work of many people and the support of many organizations.

We thank first the National Science Foundation, without whose support this conference could not have taken place.

We thank the many speakers from the United States and abroad, who shared their thoughts and expertise with us, and the conference participants who contributed to the conversation and ambience of the meeting.

The chapters in this volume were reviewed by Barbara Reys, Jeff Shih, Amanda Thomas, and the volume editors, Meg Bates and Zalman Usiskin.

Finally, we wish to thank Information Age Publishing for making this work available to the wider mathematics education community. Proceedings from the first and second international conferences are also available from Information Age Publishing (see http://www.infoagepub.com/).

CHAPTER 1

WELCOMING REMARKS

The opening session of the conference took place on November 7, 2014 in Ida Noyes Hall at the University of Chicago. The following is a transcription, with some small edits for this volume, of the welcoming remarks.

ZALMAN USISKIN

Good afternoon. It is my pleasure as chair of the program committee to welcome you on behalf of that committee to the Third International Curriculum Conference organized by the Center for the Study of Mathematics Curriculum (CSMC).

I have been asked to provide some background about the conference. Those of us associated with the center believe that curriculum is fundamentally central to what teachers and students do and learn in their study of mathematics. Our first conference examined how content decisions and textbook creation happen in countries from the highest scoring area of the world on international tests. Our second conference examined how technology and other recent developments are affecting or might affect the content of two of the main subject areas in school mathematics—algebra and geometry.

This conference has been designed to bring people from around the world to report on, and to discuss the state of the art and the future of the impact of technology on mathematics curriculum development and delivery. The idea for this conference arose from the leadership team of the center: Barbara Reys, Glenda Lappan, Chris Hirsch, Kathryn Chval, Betty Phillips, Bob Reys, Jack Smith, Steve Ziebarth, and myself.

The leadership team brought together the following people to constitute the program committee: Meg Bates, Chris Hirsch, Andy Isaacs, Barbara Reys, Jeff Shih, and Amanda Thomas.

As with the first two conferences, the financial backing has come from the National Science Foundation's support of the center and from the registrations of participants. This conference also has had the benefit of help by people working for the Center for Elementary Mathematics and Science Education here at the University of Chicago.

We are pleased that this is the largest registration of the three CSMC international conferences. Participants come from 31 states and the District of Columbia in the United States and from 11 other countries. Participants include a variety of types of personnel who work for schools in grades K–12, for universities, for research or development projects, for publishers, for grant agencies, or as independent consultants.

As with our previous conferences, we have asked individuals who we feel are among the most knowledgeable in the world to offer plenary talks. We have six plenary speakers from outside the United States, and six from inside. There are also two panels—one dealing with research on digital curricula and the other dealing with what is happening in schools that have implemented digital curricula. There are parallel sessions to enable participants and speakers to interact. These sessions are meant to be discussion sessions initiated by comments and questions from the participants.

The variety of digital materials available led us to feel that people coming to the conference might have vastly different experiences and perceptions about these materials. This led us to invite our speakers and three large publishers in the United States to demonstrate what they are doing in this regard so that we might all have some common experiences we could reference throughout the rest of this conference. Six of our speakers and two of the publishers accepted our invitation. It was an experiment, and I think it worked very well.

The speakers were asked to discuss one or more of the following topics:

- changes in the nature and creation of curricular materials available to students,
- transformations in how students learn and how they demonstrate their learning, and/or
- rethinking the role of the teacher and how students and teachers interact within a classroom and across distances from each other.

This is why we have such a long title for this conference: Mathematics Curriculum Development, Delivery, and Enactment in a Digital World.

I hope that this has given you a broad understanding of the purpose and nature of this conference. Now it gives me great pleasure to introduce Barbara Reys, the chief of the CSMC principal investigators, and the person without whom this conference would never have taken place.

BARBARA REYS

On behalf of those who work in and with the CSMC, I welcome you to this conference. I would also like to extend my sincere appreciation to Zalman and his staff at the University of Chicago. It's always a treat to come to this lovely city and to meet on this prestigious campus. It's also fitting to hold this conference at the site of the University of Chicago School Mathematics Project (UCSMP).

UCSMP has been a partner with the center since it was established in 2004 with funding from the National Science Foundation. CSMC is in its tenth and final year of funding and will complete its work in 2015.

The CSMC served the K–12 educational community over the past 10 years by focusing scholarly inquiry on the development, implementation, and use of mathematics curriculum. Major areas of work included understanding the influence and potential of mathematics curriculum standards and textbook materials; enabling teacher learning through curriculum material investigation and implementation; and building capacity for developing, implementing, and studying the impact of mathematics curriculum.

In addition to the University of Chicago, center partners include the University of Missouri, Michigan State University, Western Michigan University, and Horizon Research. Since 2004, 52 doctoral students have completed a program of study with a special focus on mathematics curriculum. In addition, over the past decade the center has published 13 books and hundreds of articles reporting on research and practical issues regarding mathematics curriculum. Our newest publication is a set of papers and talks given by Zalman Usiskin over the period of his career spanning five decades.

Curriculum remains a vibrant point of discussion among the U.S. educational community. As you may have heard, we've been debating the value of common versus localized curriculum standards. With the help of center partners and curriculum developers such as Zalman, Glenda Lappan, Chris Hirsch, and others, American teachers have also had access to cutting edge, research-based curriculum materials that focus on the development of student thinking and reasoning.

Today, a major impetus for curriculum change is the movement to electronic delivery of curriculum materials. This new format has the potential to improve learning opportunities for students. However, those of us who've been around a while know that change doesn't come easily. Too often many are wowed by the medium of technology rather than the substance of the material conveyed by technology.

It is my hope that, as a result of this conference, we will renew our commitment to high quality mathematics instruction and that we will champion use of technology to deliver curriculum that engages and supports students in learning important mathematics.

MARTIN GARTZMAN

Welcome to the University of Chicago. I am Marty Gartzman, the executive director of the Center for Elementary Mathematics and Science Education here at the university and chair of the local arrangements committee for this conference.

The University of Chicago has a proud tradition of inquiry, of learning from the best scholars and doers in the world, and of tackling big issues facing our society. The university's groundbreaking work in K–12 education dates back to the days of John Dewey, and notably includes a tremendous body of work over the past three decades by the UCSMP. Included in that body of work are a series of groundbreaking conferences—the first in 1985—that ushered in and contributed to a major era of standards-based, mathematics curriculum reform. Zal mentioned the three CSMC-sponsored conferences that have been hosted here; those were preceded by four other international conferences sponsored by UCSMP that established the model for this weekend's event. Those conferences also provided the spark for many collaborations among those who attended. We hope this conference will similarly catalyze new ideas and spawn new collaborations among those participating this weekend.

PART I
CREATING DIGITAL CURRICULUM

CHAPTER 2

DESIGNING CURRICULUM FOR DIGITAL MIDDLE GRADES MATHEMATICS

Personalized Learning Ecologies[1]

Jere Confrey

> Question: How does Apple get the balance right between design and function?
>
> We see design as everything. The beautiful product that doesn't work very well is ugly. I can't really compartmentalize. The best design we have done is those that are completely harmonious.
>
> —Jonathan Ive, Senior Vice President of Design, Apple Computer (Ive, 2014)

From July 2011 to June 2013, I led the design of a middle grades digital mathematics curriculum. In this chapter, I report on the design principles and processes used to develop that curriculum and provide examples of each element. In addition, I describe the design process and how students (and some teachers) were involved, and how the interdisciplinary structure of the team and the on-site research informed design.

The writing team consisted of 5 to 7 writers, 3 outside consultants, and an editor. A tools coordinator linked the curricular work to the engineering of the digital tools (i.e., grapher, spreadsheet). A team of three people and a psychometrician consultant worked on practice items and assessment. The engineering team was comprised of 5 to 10 engineers (depending on the scope of the work at any given time). Two UX (user experience)/UI (user interface) people and one visual designer also worked full-time with the team. Most project managers were located at the central office along with a media team.

CLARIFICATION IN LANGUAGE USE

Digital environments offer new means to address issues that have surfaced in education year after year. One set of issues involves responding to individual student differences using a variety of terms, such as differentiation, grouping, adaptivity, personalization, customization, and individualization. I begin by describing the framework I use to define these terms.

By *differentiation*, I refer to methods of instruction within a class designed to meet the diverse needs of learners. Two approaches to differentiation are *grouping* and *adaptivity*. *Grouping* can be achieved with various arrangements such as homogeneous or heterogeneous grouping. A group can be maintained over different time periods (all year, for an entire unit, by problem or problem type). Using groups to teach the same material to all students is not typically considered differentiation unless the instructor's expectations for the groups differ. *Adaptivity* implies that a product's differentiation approach is designed to automatically adjust based on prior performance levels. Its goal is to regulate the level, sequence, or dosage of problems based on a student's success rate. Determining optimal ways to differentiate instruction while protecting the fundamental benefits obtained through discourse and interaction is one of the most significant educational challenges of digital materials design.

I use the terms *personalization* and *customization* to express a related but different idea. Opportunities for *personalization* occur when students can undertake curricular opportunities based on personal interest, choice, or preference, in turn based on personal identity, goals, and/or sense of agency. In mathematics, this might mean that a student could choose to explore a concept such as a unit ratio in the context of a hot dog eating contest, a gear ratio, or the combination of paint used to make a particular color shade. In contrast, *customization* refers to opportunities provided to users to adjust the parameters of the product to meet their personal learning preferences as a client. This could include how I arrange my desktop to access tools, or how I curate my notebook in terms of use of different types of notational tools, font and font sizes, and language preferences, etc.

The importance of clarifying the use of these terms is that it allows one to use them in design to clarify how the body of the curriculum can be varied among different students to meet their needs. In the field broadly, I perceive a worrisome drift towards increased *individualization* (as in the online matter at http://www.khanacademy.org and http://www.assistments.org) that underestimates the importance and power of group and peer-to-peer interactions in learning mathematics. I use the term *individualization* to refer to methods in which students proceed individually through materials, largely on their own, checking answers against predetermined responses, having little interaction with other students, and reaching out to teachers on an as-needed basis. Based on the distinctions suggested here, a digital curriculum can serve a student's need for both individual and collective activity if it provides appropriate opportunities for personalization and customization while ensuring substantial peer-to-peer and teacher-to-class interactions; it depends on flexible grouping. Using differentiation within this approach would foster success and address issues of equity.

GOALS AND OBJECTIVES WITH SCOPE AND SEQUENCE

At the beginning of the process, I led the development of a document that summarized the principal characteristics of this product (see Table 2.1). Product development began with the creation of two scope-and-sequence documents based on learning trajectories and the Common Core State Standards for Mathematics or CCSSM (National Governors Association Center for Best Practices & Council of Chief State School Officers, 2010). The scope-and-sequence documents were organized into twelve units per year, each unit covering approximately three weeks. The first document covered the middle grades (6–8) standards over 3 years. The second compacted those materials into two years to allow a full first-year high school course to be taught to advanced students during 8th grade. To ensure adequate focus on the standards and complete coverage, each standard was assigned *primarily* to one unit. Each document outlined the learning trajectories for each unit. Consistent with the decision by the CCSS assessment consortia (Partnership for Assessment of Readiness for College and Careers [PARCC] and Smarter Balanced Assessment Consortium [SBAC]) to reduce emphasis on measurement, geometry, and statistics, these topics were typically sequenced to occur after testing. To counter a concern by teachers that these topics would be unacceptably neglected, each year one of these topics was placed in the fall semester. For instance, in sixth grade, the unit on statistics was placed fourth, following the units on ratio and proportion and percent, to reinforce those concepts in the context of building statistical reasoning.

TABLE 2.1 Principal Characteristics of the Product

Principal Characteristics	Description
Students are provided:	
1. Learning progression-based instruction around the CCSSM.	Scaffolded instruction based on proven empirical learning progressions focused on big ideas embedded in the CCSSM.
2. Personalized instruction updated by ongoing diagnostic assessments.	Personalized instruction based on student needs, and preferences supported by ongoing assessment to maximize human potential and development of learner-oriented identities.
3. Contemporary digital learning experiences tied to exciting career explorations.	Proven increased proficiency in digital forms of learning and new topics of study for readiness for future career opportunities.
Students become:	
4. Confident, independent, and persistent problem solvers.	Confident, independent, and persistent participation by a community of learners solving increasingly complex problems using contemporary tools of the discipline.
5. Proficient in argumentation and proof supporting further advanced study.	Interactive instruction involving rich representations tied to increasingly rigorous argumentation and proof in preparation for successful further study.
Teachers become:	
6. More empowered to coach a student to learn rather than to teach via lecture.	Real-time feedback and dynamic support empowering the teacher to interact with the students in a classroom that is structured less around lecture and focused more on engaging kids to learn.

The remainder of this chapter is organized into two major sections: (a) categories of design decisions, and (b) description of the design process and associated research. In the first, design decisions about the lesson structure, workspace, tools, and accompanying challenges are discussed. In the second, the application of a design study approach and interdisciplinary collaboration are outlined. Other elements of the product, including diagnostic assessment, practice and fluency, teacher support and math projects, are not discussed in any detail in this chapter.

PART I: CATEGORIES OF DESIGN DECISIONS

Category 1: Design Decisions on Lesson Structure

You have this really strange balance between being, it might appear to be a little bit stubborn, I like to call it resolute, but really pushing and pushing an idea but at the same time realizing it might not work, and at some point, you have to make that decision. (Ive, 2014)

When we started this process, we thought we were building a curriculum. As we proceeded, we realized that the scope of the work was both more extensive and potentially less extensive. It was more extensive in that with an instructional system that could include, for instance, the time on each portion of the lesson sequence, the curriculum would necessarily reach into pedagogy (how the material would be taught). In addition to potentially defining the use of time, a digital curriculum can define the lesson flow far more specifically than a printed textbook can. To consider how to create a curriculum and pedagogy combination, early on I proposed a set of possible activities with which we, as writers, could select and sequence lessons. My initial thought was for each class lesson to select from among quick checks, "hurdles," challenges, games, interactive demonstrations, controlled practice, complex problems, and closure. Some would be part of a daily routine (quick checks, closure) while others would be used as needed. Time was considered to be running during a class, with the technology providing feedback to teachers on the degree to which they were progressing based on the length of the class period. In the end, the variety of choices needed simplification to avoid over-scripting the class.

If a digital curriculum provides a structured lesson, then a question arises about access to what might be called "reference material." Traditional textbooks function as a source of content explanation with worked examples and assignment of student problems. Some contemporary curricula (e.g., Connected Math, Core Plus), instead, include more examples of problems posed and provide opportunities for students to work on them. The lesson itself is assumed to be carried out by the teacher with varying degrees of reliance on the text. In contrast, digital curriculum can contribute more directly to the presentation and staging of lessons. Working out what that might mean constituted much of the early work in design. The curriculum was less extensive in that we were not preparing any "reference" material.

Design Decision 1.1: *Workspaces differ from lesson flow.*

The math product followed the initial development of products by Amplify for English Language Arts (ELA) and Science. A core technology had been created for these two subjects that logged students in, kept a database of their progress across subjects, and provided a linked interaction structure for presenting the lesson. It mimicked the structure of a textbook in that it created an outlined format of units, lessons, and activities. A decision was made by company leadership for lessons to be delivered to students in the format of "cards" (digitally mimicking index cards) to create the lesson sequence. In ELA, other activities in writing were supported by providing a writing resource; in science, simulations were used to support independent student work.

By contrast, in math, I wanted students to spend the majority of their time actively doing mathematical work. For this reason, from the beginning I argued for a *mathematical workspace* where students could both access tools and create mathematical work that could later evolve into the work records and a student notebook of their own activity.

In this context, I proposed a fundamental distinction between *lesson flow* and the *student workspace*. Instead of lessons focusing on a lesson flow where students would be occasionally assigned to exit the lesson and work on an assignment using a simulation or a writing tool, for the math product I proposed that students work from a mathematical workspace with access to tools and that the lesson elements be brought into the workspace. A major distinction between these two approaches is that students would have control over their workspace, just as artists would control their palettes, or as writers their composition books, and that problems would come into the space as tasks to be addressed.

Design Decision 1.2: *Digital curricula should leverage opportunities for interactivity to support emergence of student thinking even as transactional approaches are used to deliver tasks.*

I also distinguished between *transactional* and *interactional* curriculum approaches. If a digital curriculum is designed for a teacher or the system to push out an activity or task, expecting a student to respond by submitting a response that is then evaluated, with the sequence then continuing, the curriculum is largely transactional with distinct moves. By contrast, an interactional curriculum is one that may have certain expected transactions but which supports many more spontaneous and contingent actions-and-responses or interactions. An interactional curriculum, I conjecture, is more likely to be student-centered and able to build opportunistically on emergent student ideas.

Because the core technology had been heavily linked to the idea of a lesson sequence delivered on cards and because ELA had an earlier publication date, gaining traction to revise that core to support a separate workspace and an interactional curriculum was difficult both conceptually as well as practically in establishing it as a priority in the overall product development. We were fortunately able to create what we referred to as a "tracer bullet" that demonstrated the feasibility of this approach, from start to finish of a lesson, before my departure. Due to the conceptual importance of the workspace and the access to tools, I next discuss this aspect of digital curricula.

Category 2: Design Principles About an Interactive Workspace and Tools

> So much about industrial design is the sense of service. What we do is we make tools for each other so the goal unlike fine art, it is not about developing narrative or self-expression, we try to make tools that make life a bit easier. I think that is quite noble and I like that idea. (Ive, 2014)

In this section, I discuss the design decisions that were practiced over and over as we built out new curricular units. That repetition and variations experienced within it allow me to describe this work as a set of design principles.

Design Principle 2.1: *Students need opportunities to engage in mathematical work using mathematical tools at levels appropriate to their knowledge.*

Some products (Core-Plus, Pearson) use interactive diagrams instead of mathematical tools; these diagrams are small applications that illustrate particular ideas (Confrey & Maloney, 1999). Confrey, Castro-Filho, and Maloney (1997) described three types of these (illustrations, representations, and simulations). The advantage of an interactive diagram is its typically narrow focus, which minimizes the time required for tutorials and learning to use the diagram. We used interactive diagrams for a few purposes when we envisioned a need for a simple tool.

However, from the inception of the curriculum, I committed to support students in actively doing *mathematical work*, at levels appropriate to their cognitive development. To accomplish this, the product required a workspace and a suite of tools. Selecting the array of tools comprised the first challenge as we selected among a graph-table capability, a dynamic geometry tool, a statistics tool, a spreadsheet, and a measurement tool. All these choices were influenced by the suite of tools already established for use in mathematics education, such as Geogebra (geogebra.org; Hohenwarter & Jones, 2007), Function Probe (Confrey & Maloney, 2008), Geometer's Sketchpad (Jackiw, 1995; Scher, 2000), Cabri (cabri.com; Laborde & Laborde, 2008), Fathom (Chance, Ben-Zvi, Garfield, & Medina, 2007; Key Curriculum Press, 2006), Tinker Plots (Konold, 2007; Konold & Miller, 2005), Sketch Up (sketchup.com; Chopra, Town, & Pichereau, 2012; Livingstone & Fleron, 2012), Microsoft Excel, and others. During the development process, the graphing calculator software Desmos (desmos.com/calculator) was introduced. Other tools examined included script recognition tools MyScript (myscript.com), MyScript's MathPad, and FluidMath (fluiditysoftware.com). Many of the landmark digital tools did not run in the tablet operating system, so a new engineering process had to be undertaken. The team started with a statistics tool and a combination of a graph and table.

Two advantages of using more general mathematics tools include: (a) students develop the ability to select appropriate tools (one of the CCSSM mathematical practices), and (b) students can use these tools in independent explorations. Furthermore, many software applications now provide for tutorial assistance to appear on the screen if the user's cursor (or finger, on a touch-screen) hovers over an on-screen feature; this approach helps mitigate the time costs of building skills with the tools.

A second challenge in building a suite of tools is to decide which tools to build and how they relate to each other. Questions emerged about whether to build separate table-and-graph and dynamic geometry tools, or a single tool in which the capabilities are merged. Desmos is an efficient and well-designed graphing tool, but its table capabilities were limited and it lacked the bi-directionality of Function Probe, with which one could see how changes made by direct graph actions were reflected in a table. The practical usefulness of an Excel-like spreadsheet was attractive, but the inconsistency between spreadsheet and algebraic notation was a potential detriment to meeting the goals (and time constraints) of the middle grades CCSSM. Symbol-manipulation tools presented another challenge: One could merge these into the display capability for graphs, but then lose the general focus on equations that is heavily emphasized in equation-solving in the middle grades.

We identified four major approaches for symbol manipulators. One was to create a touch-based solver that permits or restricts certain moves. In this approach, the student works within a microworld and "learns" based on successful equation solving (Dragonbox, dragonboxapp.com; AlgebraTouch, regularberry.com). The moves often seek to convey a sense of the actions of combining, operating on two sides of an equation, or distributing or factoring. A second approach we considered would be properties-based, in which students need to apply a property in order to make a move. This approach can be helpful in learning proofs and in understanding the reasoning behind a solution strategy. With increased student proficiency, however, the requirement of explicitly identifying each property and full delineation of steps would become onerous. A third approach is to learn symbolic moves using a manipulative as an aid, often in relation to the variable. The bars in the Singapore Middle Grades materials, and some of the features of Dragonbox, are examples of this approach. Finally, a fourth approach is a semantic calculator, in which the phrasing of a mathematical problem is transformed into a means to create and solve a hybrid equation. Each approach has virtues and no final decision was made before my departure. One process description is, however, worth mentioning in this regard. Every week, a team demo was planned, primarily by the tools coordinator. This kept the team up to date with new developments and possible design features.

Explorations that the team conducted with tools provided an opportunity to describe a first set of design principles for a product that prioritizes mathematical tools and a workspace.

Design Principle 2.2: *Notating mathematical ideas symbolically should become as intuitive as writing and sketching in traditional media.*

Inputting mathematical notation is one of the most difficult challenges for the tablet, on which screen space is at a premium and for which touch screens offer a tactile experience for manipulation of screen objects, which can serve as its own form of mathematical expression. Fingertip movements typically produce script that is too thick for mathematical notation on a small screen, while styli are easy to lose. Inputting using a keypad, whether alphanumeric or numeric, is unsatisfying as it uses up a lot of screen space, and often blocks the view of other essential features and placement. A second option is to use handwriting recognition software that transforms handwriting to notation, as does Fluid Math. Combining such a tool with an appropriate symbolic manipulator, with which students can learn (either by using a dynamic manipulative or attempting proof level justification as an option), will represent a real solution to this challenging problem.

Design Principle 2.3: *A workspace for deploying and using tools separated from lesson flow allows students to become independent and proficient tool users.*

In addition to the question of which tools to select is the question of what the characteristics of the workspace need to be. A primary corollary of my commitment to support students to act like mathematicians was to provide them a space in which they could create and curate their own compositions. Reflecting back, two primary issues arose that spawned numerous debates and explorations. The first was how the workspace should relate to the lesson flow, and the second was what the size and shape of the workspace should be.

I preferred a workspace that was separate from the lesson flow (which could be accessed independently and would support archiving items and work into a separate digital notebook). Navigation could be undertaken independently, either through: (a) the lesson flow structure, (b) the current workspace, or (c) the archived workspace or notebook. This variety would support in-class work as well as review or retrieval for elaboration, revision, and discussion. That is, if an archived workspace can be retrieved to become the active workspace, it could be worked again.

Thus, a student reviewing a lesson could access the work she had done. If she wanted to see something she had composed in the workspace to launch another challenge or project, she could work from there, creating a copy that would also be located in time in relation to her current landscape of work. It brings to mind the need for a careful and elegant system of tagging that allows students to more flexibly search an archived workspace.

The question about the size and shape of the workspace also required an interesting investigation and represents an open and compelling research agenda. One could imagine the workspace as an "infinite" tablet-wide space (we called this our paper towel roll model). In place of a book-like notebook, a student could have a rolling page that he could number and label as he wished. Alternatively, the space could be a region that would be extensive both left and right and up and down, such as in many graphics programs (Visio, Lucidchart, etc.). This scenario would offer various options for tiling the plane and working with it systematically. Another possibility includes the use of levels of scale in order for students to choose to elaborate on a point in proximity to the diagram. Prezi represents a tool that provides such features. Finding out what kinds of workspaces support student reasoning and yet are manageable by students organizationally over time is a robust open territory for research.

Design Principle 2.4: *Notebooks created by students allow them to curate their work, drawing from their workspaces and integrating lesson elements.*

Students can actively shift all or some of their work into the notebook as an act of curating. The notebook was conceived as a place where excerpts of work would be stored for future work curated by the student, along with definitions and formalizations provided by the teacher. This encourages active listening by students and decision-making about what should be kept. We did, however, place precise definitions of mathematical terms directly into the notebook during episodes of formalization within the lesson flow, to be certain students had access to accurate definitions. New forms of effective note-taking in this digital environment are an open research question.

The task and overall management of curation will evolve into another rich area for future research. Students will have many more ways to keep records and notes than in traditional curricula. How they do this, how they access them, and their efforts on knowledge retention and understanding will be a growing source of design opportunities. In a culture where "to google" something has become a verb, the question of how one accesses, notates, expands, and shares one's records and portfolios of work is an active and open line of innovation and research.

Design Principle 2.5: *Student workspaces should be under their own control in terms of what tools to select, display, and manipulate; resources to assist in this process, as forms of tiling, could be made available as needed or preferred.*

Another significant question is how tools deploy onto the workspace. While tablets offer very constrained screen sizes, swiping (left and right, up and down) offers a means to extend the virtual screen environment. Assisting the user in the effective use of space via *templating* the appearance of tools is one option; in watching students use multi-representational tools, it was evident that they exhibited a variety of preferences for how those varied displays should be configured. Again, an affordance of a digital environment is to be able to turn on or off those templates as an act of customization. Over-scripting a lesson and/or over-templating a lesson is a constant danger, especially for people (teachers, curriculum developers) who doubt students' abilities to self-organize.

> **Design Principle 2.6:** *Both students and teachers need varied and rich opportunities to interact in the context of learning together and participating in mathematical work as a community of learners.*

For the students, this will include the ability to broadcast work, collaborate in real time, and share and submit work samples. Our work in this area has barely begun, but it is a rich territory for future work, and will lead to a much more detailed set of design principles.

We conducted some initial studies around a problem designed to help students understand why a divided by b equals a/b in the context of sharing multiple wholes among multiple people. Based on prior work on equipartitioning (Confrey, Maloney, Nguyen, & Rupp, 2014), we built a small application to support multiple approaches. Students recorded their strategies (deal and share, split all and co-splitting) and then narrated them for other students as they replayed the strategies dynamically on the tablet. Student participation was active and spirited. During a second lesson, fair sharing using only ten splits to lead to decimals, the recording of the students' approaches failed (a network problem that gave rise to an inadvertent experiment) and students had to explain their strategies without the animations. The level of engagement diminished markedly, and post-test scores on the second items were significantly lower.

Teachers in these rich environments will need a dashboard that can support their interactions with students. This will include data on students' progress towards completion as well as their success. This could support teachers in collecting and sequencing examples of student work to foster rich discussions. It could also allow teachers to receive and respond to questions, and pose new ones, in real time.

Category 3: Design Principles in Building Challenges Within Units

> Consumers are discerning... I believe that we sense—we can't articulate why, I can't articulate why, but we can sense—when there has been care taken with a product. Just in the same way we sense carelessness. And sadly most of manu-

factured environment I think testifies to a degree of carelessness. It testifies get it built fast, make it cheap, make it look different. It is just that sort of carelessness. It is one of the things we strive to do for humanity, and it is a way we could serve, is just to take care. (Ive, 2014)

In this section, I describe how Challenge Problems were constructed, and articulate the design principles associated with them. Figure 2.1 shows the overall structure.

Design Principle 3.1: *For each unit, a structural overview was proposed, organized around a learning trajectory that guided the sequence of challenge problems taught in that unit.*

Confrey, Maloney, and their team proposed the 18 learning trajectories for K–8 mathematics that are outlined on the site www.turnonccmath.net (Confrey et al., 2011). For each learning trajectory, a proficiency matrix described the trajectory in terms of observable behaviors and outcome spaces describing the range of responses at different levels of sophistication (Confrey & Maloney, n.d.). The proficiency levels were also associated with particular standards.

In building a unit, the learning trajectory was used to create a "structural overview" of the unit. For example, for ratio, it was comprised of:

1. Creating a need for ratio.
2. Exploring equivalent ratios.

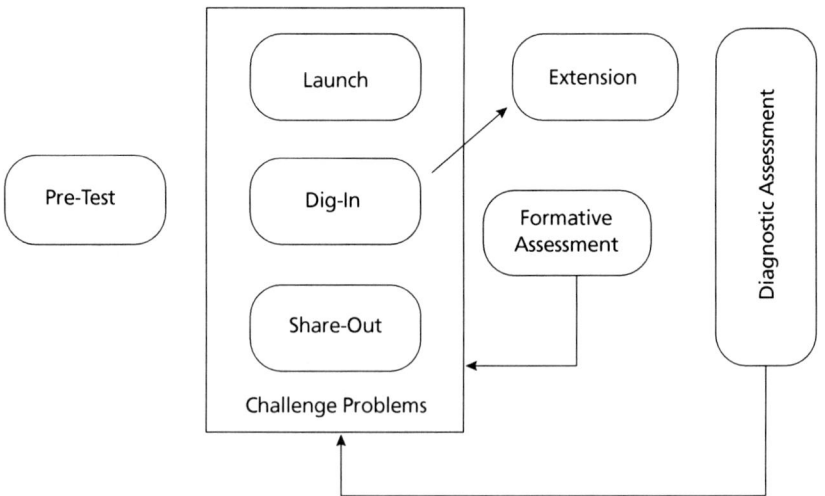

Figure 2.1 The structure of Challenge Problems within the unit structure.

3. Finding base ratios.
4. Finding and using unit ratios.
5. Building up using tables.
6. Building up using graphs.
7. Comparing ratios in graphs and tables.
8. Finding missing values by building up.
9. Finding missing values using multiplicative reasoning in ratio boxes.

A set of Challenge Problems was then created to address these steps in the learning trajectories.

Design Principle 3.2: *Students benefit from engaging with challenges that "create the need" for an idea and provide them opportunities to engage with "productive struggle."*

The design of tasks for learning is an inexact science of emerging importance (Watson & Ohtani, 2015). Many discussions of task construction begin with the mathematics concept and consider how to vary it to create tasks. For instance, the abilities defined by Krutetskii (1976)—information gathering, information processing (curtailment, generalization, flexibility, and elegance), and memory—were used as means for problem construction. Stage theories (Piaget, 1971; van Hiele, 1986) provided other approaches. Gravemeijer (2004) recounted three heuristics for instructional design from the Freudenthal Institute's program of realistic mathematics education: guided reinvention, didactical phenomenology, and emergent modeling. Based in radical constructivist and Piagetian notions, in 1991, I described "creating a need for an idea" and defined that idea in this circumstance as "a roadblock to where a child wants to be" (Confrey, 1991).

What unites all these movements has been an orientation to the student as an active constructor of his or her own knowledge. Recently, Kapur (2008, 2012) articulated a construct called "productive failure." It was based on research showing that when students are provided opportunities to invent solutions to a concept before being given a formal introduction, they learn the material more successfully. I preferred to adapt this term by describing this as "productive struggle," because students' efforts often reflect real progress towards successful solutions, rather than simply a failure to reach canonical solutions.

Examples of problems that create the need for concepts involve creating narratives to which students can relate and with which they engage. One such problem showed a video of a wind-up toy walking a certain distance in a specific number of steps. The toy was then fully wound, an action that allowed 36 steps, and placed a specific distance from the edge of a table, and students were challenged to figure out whether it would fall off the table before coming to a stop.

Design Principle 3.3: *In a challenge problem, each of the three steps (Launch, Dig-In, and Share Out) needs to do specific work.*

Inspired by reform curricula in the United States such as Connected Math, Math in Context, Interactive Math Program (IMP) and Core Plus, our Challenge Problems included a "Launch," a "Dig-In," and an "Share-Out" phase. The *Launch* phase invites the students into the problem and asks them to think about how they would go about solving the challenge problem. An important part of the launch is to give students an opportunity to interpret the problem before trying to generate possible strategies. Launches often include a video introduction to the problem. If students spend time understanding the problem (Pólya, 1957), then they are more likely to generate strategies that are productive.

Dig-In is the phase in which students offer strategies and representations to address the problem from the launch. A good challenge has multiple paths of entry, so that most students can find a toehold on the problem. It may require them to create a representation, to try out some values, to build a table, or to try to explain the problem in words. Teachers need to learn to resist giving away the problems' solutions; as they observe students inventing good ideas, they will learn to trust the students, and, in fact, to treat the challenges themselves as instructional vehicles. To assist teachers in how to help students, we provided them with suggestions about how to "jump start" the students, how to recognize productive strategies and misconceptions for the specific problems, what to emphasize, and what to delay until later.

For example, in the unit on ratios, students begin by viewing two funny video clips and participating in a poll to see which is preferred and by whom. Based on analyzing a variety of their own characteristics, the students are led to examine differences in teams' preferences, which creates the need to know how to compare ratios. How to do this is then left as an open question. This challenge is followed by another challenge in the context of making lemonade—to make more or less but make it taste the same (Noelting, 1980). Slowly students build their understanding until they are ready to return and solve the comparison successfully.

Share-Out, the final phase, is discussed further in the next design principle.

Design Principle 3.4: *When engaging with challenges, students often benefit from working with others and sharing their ideas in environments with supportive socio-cultural norms of collaboration.*

Learning to work with others with the aim of generating useful strategies is a learned skill, strongly influenced by norms created and supported in the classroom context. In the Dig-In phase, students work together and are provided a set of assets for solving the problem. Tools may be pre-set, for example, a table with headers labeled and data already entered. Access to the suite of tools is always available to support other approaches. A major research area is to decide how to support collaboration. Should each student work the problem individually or should students be provided a shared workspace? Can they enter values into the same table, or do they share their tables when they feel ready? Many tools in the professional work environment (e.g., Google docs, Lucid charts) support such flexible collaborative modes.

Because students enter Challenge Problems with different levels of preparation and different rates and paths of learning, our materials also provided students with access to an extension problem, directly related to the problem setting at hand, but incorporating further elaboration. This design permitted all students to use the time devoted to Dig-In fully, without leaving some students behind.

During the Share-Out phase, students should be able to easily share their work. Approaches that support this can vary from having students broadcast their solutions to having them record and narrate their work. If support is provided for teachers to view student submissions, then teachers can decide how to sequence student presentations for effective class discussion. In this phase, the teacher should have clear objectives about what the students should take away from the challenge. One resource for this was the delineation of a set of formalizations that the teacher could "push out" to students for discussion and for direct placement for accuracy in the notebook.

After the Share-Out, students would be given a formative assessment problem. These were quick direct checks for understanding. Feedback on students' performance on the formative assessments determined if the students were ready to undertake practice or work on a mathematical project, or instead would need more structured instruction—called an *interactive demonstration*. After 2 or 3 Challenge Problems, students would take a diagnostic assessment to see if they had successfully learned the material. The diagnostic assessment part of the project was based on the learning trajectories.

PART II:
CATEGORIES OF A CURRICULUM DESIGN PROCESS

> I feel so absurdly lucky to be part of the creative process where one day, on Tuesday, there is no idea, we don't know what we're going to do, there is nothing. And then on Wednesday, there is an idea. Invariably the idea is a thought that becomes a conversation. So the way we design, to start with, is to talk.

And it's very exclusive; it only involves a few people. And a remarkable thing happens in the process; it's the point in the process where there is the greatest change; it's when we give form to an abstract idea. (Ive, 2014)

Category 4: Design Studies in Curriculum Design

Developing new forms of curricula requires processes for designing, testing, and revising materials (many of which precede the more commonly recognized activities of pilot studies with first adopters), and field testing. Within software development traditions, *playtesting* is the preferred term for testing early phases of software. The term derives from the gaming industry where the focus is on observing users to see what choices they make, where they get stuck, and how to keep them engaged in the game. The curriculum team used playtesting to test initial designs for tools, containers, and other forms of interface design. UX/UI designers would sit with students, observe their actions, and ask them to explain their thinking. To some degree this form of testing involves investigating construct validity. UX/UI designers have an entire terminology for the kinds of behaviors they observe in playtesting.

After playtesting, I developed an application of design studies for the context of curriculum development. This entailed bringing groups of students on a regular and ongoing basis to the office site to try out Challenge Problems, Math Projects, diagnostic assessments, and practice problems. Design studies were selected for their applicability to the setting. In Cobb, Confrey, diSessa, Lehrer, and Schauble (2003), we identified five common characteristics for design studies:

1. ecologically valid and practice-oriented (focused on learning and the means to support it),
2. interventionist,
3. generative of instructional theory while simultaneously placing such theories in harm's way,
4. iterative, and
5. concerned with domain-specific learning processes, and accountable to the design process.

As we invent new approaches to digital curricular design, digital curriculum development is highly *interventionist*. We worked with students at the relevant grade levels in classroom-like settings of tables, a smart board, and a single teacher, and ran cycles of studies in 4–6 week periods with clear goals about how the materials should function. Typically, a group of homeschooled students would attend for two hours in the morning, and a group of public school students would come after school. Summer programs ran longer, usually 9 a.m. to 12 p.m. daily for 1–3 weeks. The setting was ecologically valid in that a group of students attended consistently for the duration of the study. Design studies were also accompanied by ongoing playtesting of our mathematical projects, to inform other parts of the design process.

The students would complete a pretest, with their work and discourse documented throughout the study, and take a posttest. Our pretest consisted of two categories of items, readiness items and items affiliated with the higher levels of the trajectory that would be targets for the unit. The posttest examined only the taught material.

These design studies permitted us to evaluate the tasks to see if they (a) were at the right level of difficulty, (b) elicited the kinds of student thinking we expected (or if other ideas emerged that were equally compelling), (c) engaged large numbers of grade-level students with varied experiences and backgrounds, (d) could be tackled within the time frame, and (e) made good use of the media and features. For example, one early ratio lesson was designed for students to learn to visualize ratio in the context of the length and width of a rectangle found in the room. Looking at the size of a rectangle in relation to the length of its sides was more difficult for the students than we had anticipated. Students interpreted the task to be asking for two different length objects, or to be looking at the frame of an object. Upon encountering this hurdle, one teacher move was to talk about the ratio of the heights of two people, for instance a basketball player and a student. It seemed easier for them to visualize ratio in relation to comparing the same dimension of two objects before making the internal comparison of height to width in one rectangle. It may have been the personalization and context being more concrete, the directionality of the comparison lengths being parallel, the fact that each length was in a different object or the fact that the students did not make the rectangle for themselves. For most, but not all students, abstracting the relation as parts of features of one object, for example a brick, was a premature request. Conducting the design experiment allowed us to revise and position the idea better.

Design study provided the kinds of classroom support and forms of activity that permitted us to examine the interplay among the conjectured learning trajectory, the classroom use of the materials, and the expected degree and depth of learning. The sessions were observed by researchers, the relevant writers, UX/UI designers, and typically some software engineers.

Instruction would typically be carried out by one of the writers (all of whom had teaching experience at the middle or high school level). It took considerable time to teach the writers to be investigative in their instruction—to leave room to hear student ideas. As experienced teachers, they tended to try to "pull off" the instruction, rather than listen to students enough to see where the materials created opportunities for *instructional theory generation*. At first, the instructors missed many opportunities to allow students to provide important insight into the materials; therefore, initially, much of the substance of the design study occurred during the post-class debriefing sessions. Gradually the teams learned more about how to conduct the design studies, testing conjectures in real time.

An example of a mini-investigation involved how we extended our understanding of the role of diagnostic assessments based on this kind of investigation. After we administered the first diagnostic assessment, we split students into two groups based on their performance. But we had not previously explained to the students that a differentiated set of outcomes would occur, so a substantial portion of the students did not try. After a few rounds of using this differentiated approach, we developed variations in the routines, which we explained more clearly to the students. Once the students understood the routine, they strived to do well on the assessments, and showed signs of greater appetite and respect for feedback and accountability. Many of our students came with weak preparation and doubted their own abilities; for those students, this change to internalizing the value of assessment as personal feedback was a significant accomplishment; we studied how to strengthen these elements in the product design. Furthermore, the principle of giving students rapid, automated feedback, early in the process, became standard.

As the digital technology was developed, the research questions matured to focus on linking the design of the content/task discussions with the technology affordances. For instance, in our study on the tracer bullet that combined a delivery of the content into the workspace with tools available and a means to save work to a notebook, the overall research questions were:

- Did the lesson proceed through the phases/components (Launch, Dig-In, and Share-Out) as expected?
- Where in the phase/component were the students confused about what to do or what something meant?
- Were students supported adequately by the workspace, tools, broadcasting, and container design?
- Were teachers supported adequately by the workspace, tools, broadcasting, and container design?
- Where did things break down digitally?
- How does the technology enhance the phases of the challenge?
- How do students' behaviors during Challenges change over time as they become increasingly familiar with the Challenges?

Before the design study began we would articulate the conjectures for the work, prepare the planned materials, and review them with the staff members who would teach the lessons. Team members took observational notes during the sessions; many sessions were video-recorded. Each day, the writing team would meet to debrief on the session, and clarify what had been observed. Categories of reporting included: the curriculum, the technology, the group interactions, a teacher perspective, and an examination of student work. The staff researcher and writers kept notes on the debriefing. Student work was collected and examined in revising the materials. Then the researcher would record observations and conduct focus groups with students. The curriculum writers revised materials in light of the debriefings and results. If the changes were modest, the material would not be re-tested; if the lesson failed, another version would soon be retested. Thus, a design study permitted us to create an *iterative* testing process early in curriculum development.

Applying a design study method to digital curricular development required some modifications from typical practice. First, curriculum is inherently cumulative if it is built to achieve coherence (Schmidt, Houang, & Cogan, 2002; Stevenson & Stigler, 1992). The ideas in one unit connect to those in previous units and the use of tools becomes increasingly sophisticated. But it was quite impossible for the development process by a small team such as ours to keep pace with the required learning for a group of students over a year's time. Therefore, with the exception of the first unit, most units were tested without the students benefitting from the cumulative effects of prior units. The design studies were further complicated by the fact that many students lacked adequate readiness for the curriculum. During a period of transition to CCSSM, this weakness is likely to be most severe, but it is also encountered in any unit that has topics from an earlier unit as a prerequisite.

The ongoing presence of students in our lab helped the entire team keep their focus on serving the target audience, namely, students. It also promoted a sense of community among the teams. Most software engineers and even UX/UI designers in this field have limited experience in working with students, and their work profited from the ongoing interactions.

Category 5: Interdisciplinary Activity

A major difference in building digital curricula is the extent of interdisciplinarity of the teams. Our design teams consisted of writers, media producers, UX/UI experts, visual designers, and software engineers. Learning to communicate successfully across the teams took time and considerable, iterative effort, but it was an exciting experience.

Most curriculum writers do not know what UX/UI experts are. UI (user interface) designers focus on how the interface works for the user. UX (user experience) experts focus on the overall experience of the user; their work in the curriculum project entailed a broad concern for how the student and the class experience the material. These experts possess invaluable knowledge about how to make full use of, for example, the gesture set for a tablet (i.e., possible ways of touching the screen to interact with a tablet), and they have a sharp eye for recognizing when users experience information overload, inconsistency in design, or a lack of clarity or intuition about how the interface should operate. They are concerned with the physical and technical issues surrounding input and output of the "device" (tablet, laptop, etc.). They often serve the role of communicating with the engineers about how certain tasks can be accomplished and communicating options back to the design team. The user experience certainly involves interface use, but the UX designer is concerned with the entire experience of the user—from initially picking up the tablet, working with it over time, and responding to the experience as a whole. Learning to work with UI/UX designers requires a traditionally-trained curriculum writer to learn a whole new set of terminology and understanding about their aesthetic as to what the experience of the student and teacher should be.

UI/UX designers typically start with what are called *user stories*. User stories are comprised of a list of what a user wants to be able to do or not do. The statements are typically of the form: "As a user, I want to know what the lesson is about." These user stories ensure that the range of needs of a user is addressed in designs. User stories are important in the design of curriculum because they open up the variety of ways students might interact with curricular material.

From user stories, the UI/UX team builds *wireframes*, which describe the sequence of how the software will work from one phase to the next. Relying on the user story to indicate what needs must be met, a wireframe shows how this is realized in the interface and provides an opportunity for the team to discuss alternatives for how the students and teachers will interact with the product. Software engineers work from the wireframes to build the product.

Digital software development also includes *visual designers* and *media experts*. Visual designers make decisions about the color palette, choice of fonts, and consistent arrangements as well as the visual assets used. Media experts find the digital photos or draw the assets; for video, they write the scripts and build the video or animations.

A major challenge in math curriculum development is how the accuracy of the materials is maintained through these processes of bringing proposed curricular tasks to life. It is rare to find designers (UI/UX, visual, media) who are well-versed in mathematical thinking; unintentional errors easily end up in the materials. At a subtler level, preserving the challenge in a problem is an even more nuanced skill. In an audio or print script, a small change in wording or an incorrect intonation can easily lead to dilution, misdirection, or loss of cognitive effectiveness of a challenge for a student. But these are potential problems that an interdisciplinary team must simply learn are lurking, and must work together to evaluate at each step of the creative and development process, and for which the mathematics educator must be especially vigilant from project initiation to publication.

The benefits of this kind of interdisciplinary work cannot be underestimated. Many creative people inhabit these various disciplinary communities—they offer new opportunities in curriculum development. They understand narrative and drama and have a lot to offer about how to get students' attention and keep them engaged.

Category 6: Applying Agile Methodology to Curriculum Design

Changes in how to develop curriculum in a digital environment flowed in both directions in the project. The software teams were accustomed to using agile methodologies, in which the development of software is undertaken through shared responsibility and decision-making within a managed process of prioritization. Based on this approach, we developed an application of agile methods to curriculum development.

Agile processes define methods to achieve the desired outcome in a defined period of time. As defined in Wikipedia (n.d.),

> *Agile software development* is a group of software development methods in which requirements and solutions evolve through collaboration between self-organizing, cross-functional teams. It promotes adaptive planning, evolutionary development, early delivery, and continuous improvement, and encourages rapid and flexible response to change.

The agile process consists of setting goals and then building a backlog—a list of tasks to be completed. The listing in the backlog can initially be rather vague and broad. It is the responsibility of a team leader or product manager to "groom the backlog," to constantly re-evaluate and reprioritize the goals, and to revise the task descriptions, so they are defined and can be completed in a short (commonly two weeks) time period called a *sprint*. The process was managed using project management software such as JIRA (by Atlassian, Inc.) for large-scale efforts or Trello (Trello, Inc.) for smaller ones. The development team is broken into small groups which take on particular tasks from the backlog. Every two weeks, during sprint planning, team members select the tasks they will work on. During a daily (or less often) *scrum*, groups report on their progress. This process of continuous development ensures that teams are self-organizing within reasonable goal and task articulation and that the process itself is flexible, and can be adapted as needed.

Applying agile methods to curricular design and the writing teams' work is an interesting challenge, in part due to the sequential dependencies of the later curricular material on earlier units. Teams can be focused on different units, but a considerable level of coordination among units (and therefore among teams) is required. To build the curriculum units, we created unit overviews based on the learning trajectory work licensed from North Carolina State University (Confrey et al., 2011). From these outlines, decisions were made as to how many challenges would be developed and on what topics they would be written. The agile methodology then applied to how a team would create a challenge, how they would playtest it and use it in a design study, and how they would propose media elements and the use of tools or interactive diagrams. Again the use of a management software ensured that drafting, testing, revising, reviewing, and editing would be conducted in a coordinated and systematic way.

CONCLUSIONS

Designing, developing, and refining digital curricula require new forms of authoring. An authoring team will consist of people with diverse perspectives and skills. Nonetheless, the major challenge in this area, where students and teachers are likely to participate in actively choosing materials, will be to ensure that students have a coherent experience. Coherence will need to be defined in this context to productively balance the experience of the student with the character and structure of the discipline.

After two years of building and leading a well-supported team to do this work, I draw some tentative conclusions on content, tools, assessment, and processes, and offer some of these as a summary of the points in the chapter:

- Using tablets (mobile technologies) and digital, active media excites students. They are enthralled by the richness of visual presentation, the inclusion of narrative elements, the dynamic displays, and the power of interactivity and feedback with and from others and the computer.
- Students need to work with tablets in a heads-up, not heads-down, pattern. For a curriculum to be student-centered, students must be the ones actually doing the mathematical work as communities of learners, and must be trusted to handle complexity (emphasis on interactional instead of transactional approaches within the curriculum).
- Challenges, explorations, projects, or investigations are effective ways to engage students and provide them with opportunities to invest in the ideas and to experience productive struggle. Tied to proven issues in learning and carefully sequenced, these resources

can form the background of a strong curriculum; however, they take longer in classrooms than one anticipates.
- Collaborative work with presentations helps students' confidence and persistence. Providing students access to rubrics for projects helps them learn to self-regulate. Rubrics need to be content-specific and must avoid giving away solutions. Teachers also need to learn to use complex problems effectively and to scaffold students without giving away the key ideas. Learning to support projects occurs over a longer time span.
- Student readiness at grade level for new materials associated with CCSSM standards is an ongoing challenge in the development of new materials that begin with middle grades.
- Tools require instructional time. Although this may slow down initial progress in a curriculum, the investments of time and effort pay off over time as students reach the more sophisticated levels of use. Using tools well is a major part of supporting mathematical work (and of supporting career readiness).
- Diagnostic assessment based on learning trajectories seems to be an effective way to improve success. Students respond well to instantaneous feedback, and appreciate opportunities to decrease the risk of wrong answers. When they understand the assessment system and know the implications of success or failure, they learn to use and appreciate the multiple opportunities provided to learn the material.
- Making full use of options within the media (e.g., video, gesture, and social media) requires a creative, diverse team that includes script writers, game makers, and clever UI (user interface) designers. Productive teams must include deep, diverse content expertise and willingness to trust student narrative sense. A content expert should have the ultimate creative control of the final decisions for mathematical correctness and consistency.
- Digital curriculum designs require numerous iterations, and content experts must be involved continuously to ensure that content elements do not deteriorate or descend into errors or trivial examples. Design and media teams must continually challenge the content experts to incorporate strong use of narrative to relate to student worlds and generate multiple avenues to engagement.
- Because time is the most precious resource in the classroom, curriculum development teams must take great care not to expect too much too quickly. The digital environments must be sufficiently simple and elegant to be supportive and not overwhelm students and teachers. Teams must also beware that responding to perceived time pressure can easily lead to teacher scripting, which diminishes the opportunities for interactivity and emergence of student approaches.

- It is critical to build digital forms of the resources as early as possible. Early prototyping with paper and pencil can appear to reduce development costs, but remaining in pencil and paper mode longer, instead of getting designs into digital form sooner, will constrain innovation that leverages the affordances of the technology. The more iterations of design, development, and testing with the digital forms, the more the final product is likely to be robust, flexible, adaptable, student-centered, and interactional.
- Testing that involves students and teachers should be varied and frequent, with clear goals for each type of testing: The result will be curriculum revision cycles that are quicker and more responsive to real time data. Ultimately, after significant numbers of students are using the materials, large-scale data analysis will further inform design, but only if design has been well understood going into the process.
- The design principles in this chapter are only one possibility.

The fundamental message from this work is that research has a crucial role in the development of digital curricula. Clear theoretical distinctions and categories of activities, informed by deep knowledge of existing research, are needed to drive student-centered innovation in curriculum. As we designed and developed the materials, we constantly drew upon knowledge produced by the mathematics education and larger learning sciences communities.

In addition, the move to digital worlds opens up many new avenues for research, not just *what* to study but *how* to study it. Defining appropriate levels of personalization while achieving coherence in student experience and socio-technical participation patterns is the next great challenge. Scholars must insist on a variety of high quality research efforts in curriculum development and implementation, including:

- careful application of design studies during development;
- iterative processes for improvement, based on analytics; and
- implementation and outcome measures that can support deeper understanding of curricular effectiveness.

These research methods *must* be distinguished from, and independent of, market research. Market research is designed to determine what the consumer is looking for and is likely to purchase. However, this alone cannot define how to build new approaches. What is needed, instead, are new approaches that build on what students will find engaging and define what new practices of teaching and learning look like, building from what we know about student thinking. To ensure the highest quality mathematics and pedagogical approaches, research and development teams also need to include both mathematics education researchers and expert teachers. Only then will our digital materials accomplish their intended purposes.

NOTE

1. This chapter reports on the work undertaken from 2012 to 2014 under my leadership, focusing on how the products built relate to what we know from research. It also outlines the processes built to design and revise digital curricula. I am no longer directly associated with Amplify, Inc. (due to creative differences) except as the primary author of Math Projects. Nothing reported here necessarily represents the opinions of Amplify.

REFERENCES

Chance, B., Ben-Zvi, D., Garfield, J., & Medina, E. (2007). The role of technology in improving student learning of statistics. *Technology Innovations in Statistics Education, 1*(1), 1–26.

Chopra, A., Town, L., & Pichereau, C. (2012). *Introduction to Google SketchUp*. Hoboken, NJ: John Wiley & Sons.

Cobb, P., Confrey, J., diSessa, A. A., Lehrer, R., & Schauble, L. (2003). Design experiments in educational research. *Educational Researcher, 32*(1), 9–13.

Confrey, J. (1991). Learning to listen: A student's understanding of powers of ten. In E. von Glasersfeld (Ed.), *Radical constructivism in mathematics education* (pp. 111–138). Dordrecht, Netherlands: Kluwer Academic.

Confrey, J., Castro-Filho, J., & Maloney, A. (1997). Interactive diagrams: A new learning tool. In J. A. Dossey, J. O. Swafford, M. Parmantie, & A. E. Dossey (Eds.), *Proceedings of the nineteenth annual meeting of the North American chapter of the international group for the psychology of mathematics education: Vol. 2.* (pp. 579–584). Columbus, OH: ERIC Clearinghouse for Science, Mathematics, and Environmental Education.

Confrey, J., & Maloney, A. (n.d.). *Proficiency matrices for learning trajectories for the common core state standards for mathematics*. Unpublished manuscript.

Confrey, J., & Maloney, A. (1999). *Interactive diagrams*. New York, NY: McGraw-Hill.

Confrey, J., & Maloney, A. (2008). Research-design interactions in building Function Probe software. In M. K. Heid & G. Blume (Eds.), *Research on technology and the teaching and learning of mathematics. Vol. 2. Cases and perspectives* (pp. 183–210). Charlotte, NC: Information Age.

Confrey, J., Maloney, A. P., Nguyen, K. H., & Rupp, A. A. (2014). Equipartitioning, a foundation for rational number reasoning: Elucidation of a learning trajectory. In A. P. Maloney, J. Confrey, & K. H. Nguyen (Eds.), *Learning over time: Learning trajectories in mathematics education* (pp. 61–96). Charlotte, NC: Information Age.

Confrey, J., Nguyen, K. H., Lee, K., Panorkou, N., Corley, A. K., & Maloney, A. P. (2011). *Turn-on common core math: Learning trajectories for the common core state standards for mathematics.* Raleigh, NC: North Carolina State University. www.turnonccmath.net

Gravemeijer, K. (2004). *Creating opportunities for students to reinvent mathematics.* Regular lecture at the 10th International Congress on Mathematical Education (ICME-10), Copenhagen, Denmark, July 7, 2004.

Hohenwarter, M., & Jones, K. (2007). Ways of linking geometry and algebra: The case of geogebra. In D. Küchemann (Ed.), *Proceedings of the British society for research into learning mathematics, 27*(3), 126–131.

Ive, J. (2014). *Apple's Jonathan Ive in conversation with Vanity Fair's Graydon Carter.* New Establishment Summit. San Francisco, 10/16/2014. Retrieved from http://video.vanityfair.com/watch/the-new-establishment-summit-apples-jonathan-ive-in-conversation-with-vf-graydon-carter

Jackiw, N. (1995). *The Geometer's Sketchpad.* [Computer software]. Emeryville, CA: Key Curriculum Press.

Kapur, M. (2008). Productive failure. *Cognition and Instruction, 26*(3), 379–424.

Kapur, M. (2012). Productive failure in learning the concept of variance. *Instructional Science, 40*(4), 651–672.

Key Curriculum Press. (2006). *Fathom Dynamic Data™ Software* (Version 2.03) [Computer software]. Emeryville, CA: Key Curriculum Press. Available: http://www.keypress.com/x5656.xml

Konold, C. (2007). Designing a data analysis tool for learners. In M. Lovett & P. Shah (Eds.), *Thinking with data: The 33rd annual Carnegie symposium on cognition* (pp. 267–292). Hillside, NJ: Lawrence Erlbaum.

Konold, C., & Miller, C. D. (2005). *TinkerPlots: Dynamic data exploration.* Emeryville, CA: Key Curriculum.

Krutetskii, V. A. (1976). *The psychology of mathematical abilities in schoolchildren.* Chicago, IL: University of Chicago Press.

Laborde, C., & Laborde, J.-M. (2008). The development of a dynamical geometry environment: Cabri-Géomètre. In M. K. Heid & G. Blume (Eds.), *Research on technology and the teaching and learning of mathematics. Vol. 2. Cases and perspectives* (pp. 31–52). Greenwich, CT: Information Age.

Livingstone, J., & Fleron, J. F. (2012). Exploring three-dimensional worlds using Google SketchUp. *Mathematics Teacher, 105*(6), 469–473.

National Governors Association Center for Best Practices & Council of Chief State School Officers. (2010). *Common Core State Standards for Mathematics.* Washington, DC: Authors.

Noelting, G. (1980). The development of proportional reasoning and the ratio concept part I: Differentiation of stages. *Educational Studies in Mathematics, 11*(2), 217–253.

Piaget, J. (1971). *Genetic epistemology.* New York, NY: W. W. Norton.

Pólya, G. (1957). *How to solve it.* Garden City, NY: Doubleday.

Scher, D. (2000). Lifting the curtain: The evolution of Geometer's Sketchpad. *The Mathematics Educator, 10*(1), 42–48.

Schmidt, W., Houang, R., & Cogan, L. (2002). A coherent curriculum: The case of mathematics. *American Educator,* 1–18.

Stevenson, H. W., & Stigler, J. W. (1992). *The learning gap: Why our schools are failing, and what we can learn from Japanese and Chinese education.* New York, NY: Summit Books.

van Hiele, P. M. (1986). *Structure and insight: A theory of mathematics education.* Orlando, FL: Academic Press.

Watson, A., & Ohtani, M. (Eds.) (2015). *Task design in mathematics education: An ICME study 22.* Heidelberg, Germany: Springer.

Wikipedia. (n.d.). *Agile software development.* Retrieved from http://en.wikipedia.org/wiki/Agile_software_development

CHAPTER 3

DEVELOPING AND IMPLEMENTING "SMART" MATHEMATICS TEXTBOOKS IN KOREA

Issues and Challenges

Hee-chan Lew

INTRODUCTION

"SMART," an abbreviation for "self-directed, motivated, adaptive, resource-enriched, technology-embedded," is a word to describe a new kind of educational system recently proposed by the Korean government (Ministry of Education, Science, and Technology [MEST], 2011a). It signifies an intelligent, adaptive learning system to innovate the educational environment. The focus of the system is broad and includes content, methodology, and assessment for transforming the current uniform and standardized education for the industrialization era towards an integrative and custom-made education for the information era.

Digital Curricula in School Mathematics, pages 35–51
Copyright © 2016 by Information Age Publishing
All rights of reproduction in any form reserved.

Developing digital textbooks was the initial strategic move to drive forward the "smart" education reform. Other reform initiatives focused on online education, sharing various ready-made materials, and improving teacher education. The Korean government set out to complete the development of digital textbooks for science, mathematics, social studies, and English by 2013 and to have them field-experimented in 2014 (MEST, 2011a). However, in the case of mathematics, the Ministry of Education (MOE) of the new government decided to postpone the development of digital textbooks for an indefinite period because the draft version was not different from a traditional textbook in its functional aspect, a situation that resulted from the non-existence of a good authoring tool and an inexperienced sense by developers of the opportunities available with digital textbooks.

Based on this initial attempt at developing smart mathematics textbooks, this chapter discusses issues and challenges that appeared in developing and implementing a digital mathematics textbook with 11th grade students. It also introduces the development of a digital textbook in the environment of Cabri LM as an authoring tool to improve students' involvement in mathematics discussion. Appreciating the essence of mathematics education as involvement in a rich classroom environment, this chapter discusses ways to provide concrete guidance in managing classroom interaction with students. Finally, this chapter features challenges to consider in moving toward practical and widespread use of digital textbooks.

TECHNOLOGY USAGE IN MATHEMATICS CLASSROOMS

Many excellent software programs, such as LOGO, Geometer's Sketchpad (GSP), Cabri, Cabri 3D, Fathom, and EXCEL, have been introduced in mathematics education since 1980. There have been various programs in Korea for preservice and inservice teachers to learn how such software can be used in teaching mathematics. Also, a number of resources were available to teachers online. This software, together with various innovative technological devices, such as PCs, the Internet, notebooks, tablet PCs, and projectors, have the potential to change the mathematics classroom learning environment.

However, the use of technology in the Korean mathematics classroom is rare. Only a few Korean teachers are reported to have used computers in their mathematics classrooms (Jeong, 2014). In many cases, teachers used them inefficiently and unproductively (Lew & Jeong, 2014). It is distressing that high quality software and innovative tools have had so little impact on teaching practices for mathematics education in Korea.

Drijvers (2012) indicates that successful use of technology in mathematics classrooms requires consideration of several factors, including the design of the digital tool, corresponding tasks exploiting the tool's pedagogical potential, facilitation by teachers, and the educational context where the tool is used. However, in Korea, the textbook as the key medium for task selection/use might be the most critical factor as it is the material teachers use most in their classrooms. The reliance on textbooks might be a main reason that the Korean mathematics classroom environment is passive and that the teachers are conservative in their integration of technology.

Because the paper textbook is well-organized and sequenced, Korean teachers rely on it for problems/tasks and the order in which mathematics is presented. Teachers generally do not change a teaching sequence presented in the textbook or re-interpret the textbook lesson because it is easier for them to merely follow all aspects of the textbook. Traditional Korean paper textbooks are designed to teach mathematics without technology. Teachers prefer the traditional "chalk and board" to other tools, such as computers, in their classrooms. Also, the paper textbook format does not lend itself to the dynamic character of the computer. There are many examples of the use of technology within paper textbooks (for the purpose of getting approval by the MOE), but they do not induce a meaningful inquiry process based on spontaneous activities undertaken by students themselves. Computers are sometimes used by teachers to show a dynamic procedure appearing in the textbook. For example, consider the following process (Park et al., 2004) for the task: "Draw a tangent line to Circle O through point P" as shown in Figure 3.1.

The textbook does not show the reason why the midpoint M is important in drawing the tangent line PQ. The textbook leaves this discussion in the teacher's hands, but teachers usually just follow the process in the textbook without the discussion of the midpoint of the hypotenuse of the right triangle OPQ. Even when teachers show the process using a computer program like GSP or Cabri, students just "see" the process through the teacher's explanation without acting or thinking. Therefore, there is no chance for students to appreciate the process because every classroom usually has only one PC, projector and screen. "Teachers talk and show, and students see and hear" is a long tradition of mathematics education in Korea.

The activities related to technology are not generally assessed in mathematics, which means that activities with computers are not a matter of concern to students in Korea. All students have to do something rather than merely seeing, and, thus, a classroom with a paper textbook falls short of what we know students need for a good mathematics education. A dynamic textbook is needed for students to explore mathematical ideas rather than just seeing the process through a teacher's explanation.

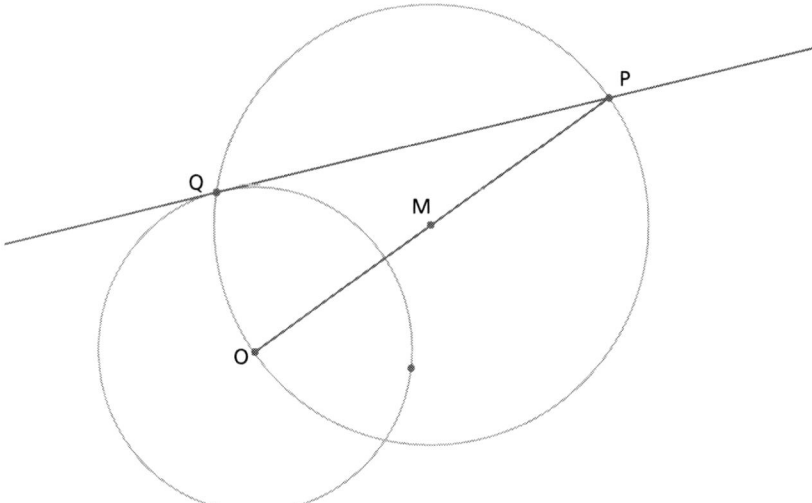

Figure 3.1 Drawing process of a tangent line introduced in the Korean textbook: (1) Draw a segment *OP*. (2) Find the midpoint *M* of *OP*. (3) Draw a circle with the radius *MO*. (4) Find the intersection point *Q*. (5) Draw a line *PQ*.

SMART BOOKS

The history of books has developed through the following four steps based on the material used: Stone/Clay (Hammurabi code of law), Papyrus/Wood (Rhind Papyrus), Parchment (Archimedean Codex), and Paper (*The Nine Chapters on the Mathematical Art*). Stone/Clay and Papyrus/Wood books have a two-dimensional space for text and Parchment and Paper books have three-dimensional space in the sense that they have a depth. Three-dimensional books can contain more information than two-dimensional books. But nowadays, we have another kind of book, the digital book. This book is conveyed via a tablet PC and is more than three-dimensional in that it can contain other kinds of information, such as sound, video clips, and links to other information.

All books are wise, but not all books are "smart." Stone/papyrus/paper books do not provide feedback that responds to readers' actions. "Feedback for learning" is similar to the "Student Programs Computer (SPC)" that Papert (1980) specified. LOGO, Cabri, and GSP have the same principle: If students act on a computer, then computers provide feedback on that action. Compared with software with an SPC mode, Computer Assisted Instruction (CAI) provides students with limited feedback.

SMART DIGITAL MATHEMATICS TEXTBOOKS

Digital textbooks might be able to provide the smart environment for students to engage in and appreciate mathematics discussion with teachers and peers based on their activities. However, they are just a starting point rather than a solution because all digital books are not necessarily smart. A "smart" digital math textbook allows for student-driven, action-based experience for both understanding concepts and exploring the problem-solving process, which is not easily embodied in a paper textbook. To be effective, the smart textbook should be the main teaching material rather than supplementary material to be used in conjunction with a paper textbook.

When the Korean government proposed "An Agenda for Action" (MEST, 2011a), the development of a digital textbook was one of the main strategic goals to drive forward smart education reform. However, in the case of mathematics it seems to have failed to reach the goal. The following figure shows one page of the digital textbook developed by the government-supported research team (Figure 3.2). This e-textbook allows one to attach a ready-made video clip, provides a solution when students click on a problem, and includes some other functions such as zooming in and out. This e-book is not necessarily smart because it provides limited on-demand feedback to students. This is just an e-textbook with solutions and with links to ready-made supplementary materials. This example raises three issues

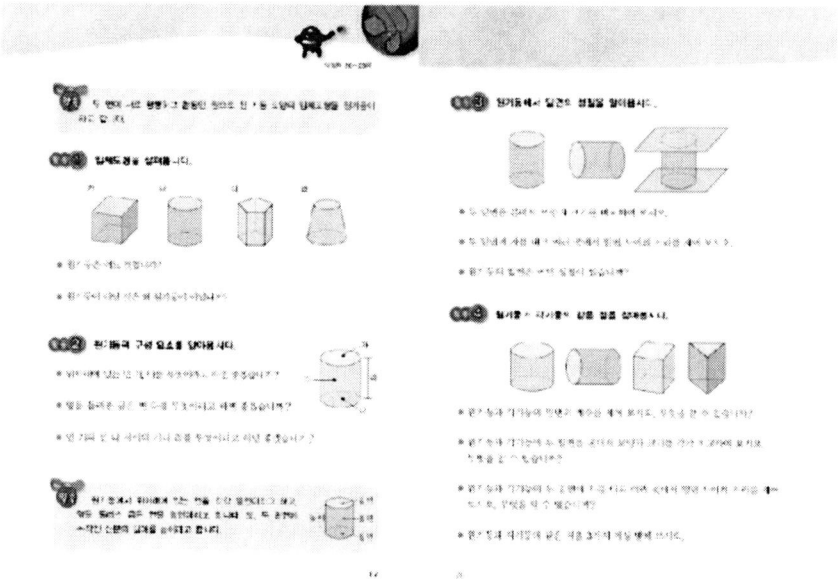

Figure 3.2 One page of a digital textbook made by the Korean government.

regarding digital textbooks: the authoring tool, the rationale for a digital textbook, and the learning environment.

Issue of the Authoring Tool

The textbook described previously was designed with SDF (Simple, Document, Flash) Studio2 as an authoring tool. It is a Flash-based e-book authoring tool developed by a private company (F-desk) and a standardized tool authorized by the Korean government. It has the advantage of strong accessibility in the sense that an e-textbook made with SDF Studio2 can be viewed on all Mac iOS, Android OS, and Windows OS and can be attached with multi-media images, video, MP3, and URL as a tool, which conforms to internal standards of the International Digital Publishing Forum and the World Wide Web Consortium. However, unlike social studies, science, or English, which do not need students' operation frequently, SDF Studio2 might not be good for mathematics because it does not support a student's spontaneous operation, which is essential for understanding and exploring mathematical concepts. A high-quality authoring tool is required to develop a high-quality digital mathematics textbook. This is a hot issue raised in the process of the development of digital textbooks in Korea.

Issue of the Rationale for a Digital Textbook

According to the MOE (1997) and the MEST (2011b), computers should be utilized to improve understanding of concepts and problem solving or thinking abilities in classrooms. Current mathematics classrooms in Korea still depend mainly on teachers' explanation and focus on low-level problem solving to apply skills and knowledge mechanically, although higher level problem solving has gradually been emphasized (MEST, 2012). To understand mathematics concepts meaningfully and increase problem solving ability, mathematics education should focus on connection between concrete and abstract ideas, which is the concept of Piagetian reflective abstraction. According to Piaget (1985), intellectual development is a continuous process of reflective abstraction that is drawn from the coordination of several actions, including awareness, analysis, and generalization, and is different from the physical abstraction which derives its sources from the physical properties of objects.

The current emphasis in mathematics education in Korea gives rise to students' poor attitudes towards mathematics as shown in TIMSS and PISA. In TIMSS 2011 (Mullis, Martin, Foy, & Arora, 2012), despite their high performance, Korean 4th grade and 8th grade students demonstrated lower levels of valuing mathematics or confidence in their knowledge and skills than students in other countries. According to the mathematics education policy issued by the Korean MEST (2012), the connections between mathematics and other subjects like science, technology, engineering, and art, which are together referred to as STEAM, are to be accentuated so that students can have a positive outlook towards mathematics.

A smart textbook provides a natural environment that encourages students to reflect on their actions and to encapsulate actions performed previously. As well, a digital textbook should create a good environment to connect mathematics to everyday life and to integrate various representations for a more interesting and meaningful mathematics education. In this environment, students could foster managerial skills for problem solving, increase intellectual curiosity, and engage in integrated thinking to connect knowledge of diverse fields.

Learning Environment Issue

To activate computer use in mathematics education, we need a new classroom learning environment where all students can explore and share ideas among each other and teachers can record and assess students' activities on computers. Technology may serve as a "Trojan mouse" to reform mathematics classrooms by changing teaching styles. The term is derived from the story of the wooden horse used to trick defenders of Troy by concealing warriors as they entered into the city in ancient Anatolia. Although technology has infiltrated classrooms successfully, nobody guarantees its successful mission because there is no warrior in the mouse, unlike the situation with the Trojan horse. To manage a dynamic textbook well, notebooks or tablet PCs for all students are required in classrooms. A "learning management system (LMS)" is also needed so that teachers in classrooms can record, monitor, and assess students' activities during classes, and students' tablet PCs can be connected with the projector for sharing their ideas with other students.

DEVELOPING A SMART MATHEMATICS TEXTBOOK

In this section, I describe some smart textbook activities for teaching the quadratic curves of the parabola, hyperbola, and ellipse. The intention is to inform the Korean MOE of issues and challenges to consider when developing a digital textbook. Cabri LM as an authoring tool was very useful in designing this dynamic mathematics textbook. In Korea, the quadratic curves are taught by the use of algebraic methods, such as equations, rather than geometric methods like construction. As a result, students do not have geometric intuition regarding the curves. The following example illustrates the cost of a problem-solving technique based solely on algebraic calculation.

Feature 1: Action-Based

This smart textbook begins with a lesson encouraging students to locate animals at points whose distances from a river and font are the same on a "physical" plane (Figure 3.3a). Students then move to a mathematical plane to let them locate the same points (Figure 3.3b). Based on this activity (showing the upper vertex of the triangle), students construct the point *P* to make the two sides of an isosceles triangle *FPH* and draw a parabola by dragging the point (the independent variable) on the directrix (Figure 3.3c). Then the definition of the parabola is introduced (Figure 3.3d). This sequence is the reverse of the way the current paper textbook in Korea develops the ideas, in which properties are taught after the definition is introduced. Then, students draw a parabola with a focus they determine and a vertex as the origin on the Descartes plane, which is a higher abstract plane than Euclidean space, and confirm and generalize its properties (Figure 3.3e). Next, students find an equation for the parabola using the properties of the parabola, which is another movement to a higher abstract level (Figure 3.3f).

Feature 2: Problem Solving Centered

The problems shown in Figure 3.4 are classified as high-level problems in the current paper textbook (Lew et al., 2009). However, if students have experience constructing points that satisfy the condition of the definition, the problems are expected to be much easier because students can use their intuition to find and enact strategies.

Figure 3.4a shows the problem to find the length of *MN* if $PQ = 10$ when the line containing the focus *F* of the given parabola meets the parabola at *P* and *Q*, *M* is the midpoint of *P* and *Q*, and *N* is the midpoint of the segment *RS* where the perpendicular lines containing *P* and *Q* meet the directrix of the parabola at *R* and *S*, respectively. Figure 3.4b shows the problem to find the centers of all inscribed circles of the given half circle with diameter *AB*, then to explain that the set of the centers of such inscribed circles makes a parabola. Figure 3.4c shows the problem to prove that if *P* is a point on the given parabola with a focus *F*, the line containing *P* and the midpoint *S* of the segment *RF* is a tangent line of the parabola, where *R* is the intersection point of the directrix ℓ and its perpendicular line passing

Developing and Implementing "Smart" Mathematics Textbooks in Korea ▪ **43**

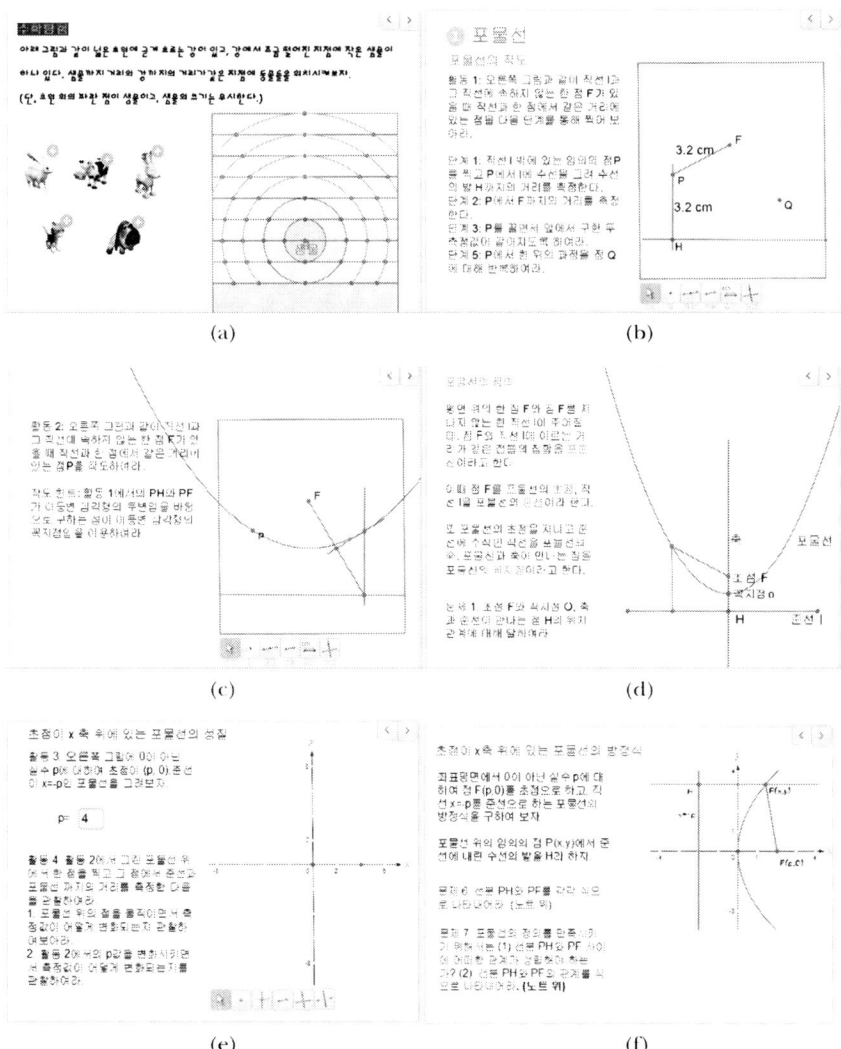

Figure 3.3 Action-based textbook.

through P. Figure 3.4d shows the problem to explain that $FP = FT$ if P is a point on the parabola with focus F and T is the intersection point of the axis of the parabola and its tangent line passing through P, and then, to prove that rays of light reflected on the parabola after radiating from its focus go out parallel to the axis of the parabola.

A digital environment where students can draw various figures on the problems precisely helps them find relations among components of the

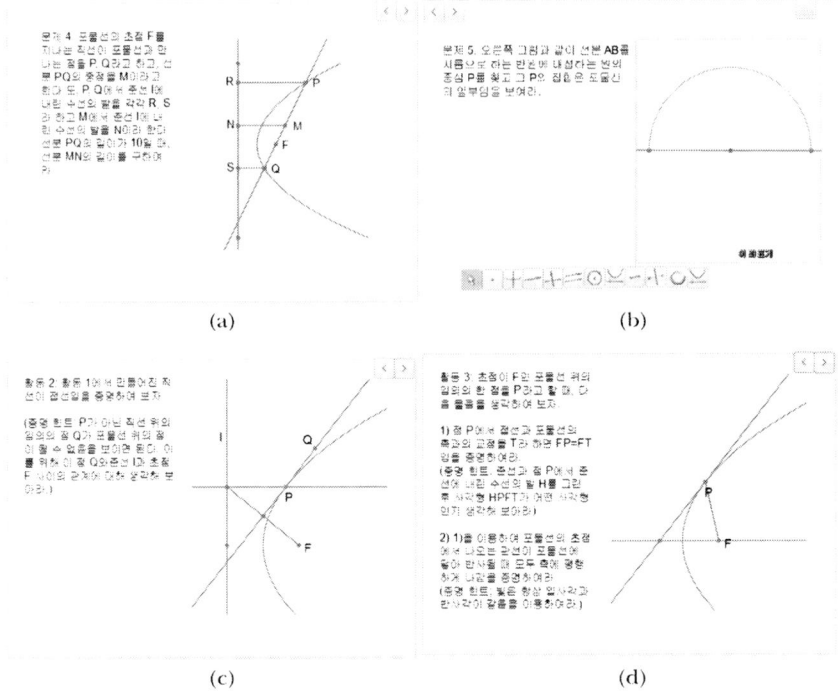

Figure 3.4 Problem-solving centered textbook.

problem situations that are difficult to find in a figure roughly drawn on paper. Furthermore, the environment allows students to perform various experiments to find relations by drawing, measuring, erasing, and manipulating figures easily as well as dynamically. In paper and pencil circumstances, it is almost impossible to perform the experiment because the figure drawn on the paper cannot be manipulated.

Feature 3: Interconnection of Mathematical Representations

Figure 3.5 shows that the three curves (parabola, ellipse, and hyperbola) are constructed by a similar methodology, namely to find the vertex point of an isosceles triangle, which demonstrates that the three quadratic curves are interconnected. Figure 3.5a shows the problem to construct a point P whose distances from the given line ℓ and the given point F, which is not on the line ℓ are the same. Figure 3.5b shows the problem to construct a point P such that when there exist the given points A and B, the sum of the lengths of PA

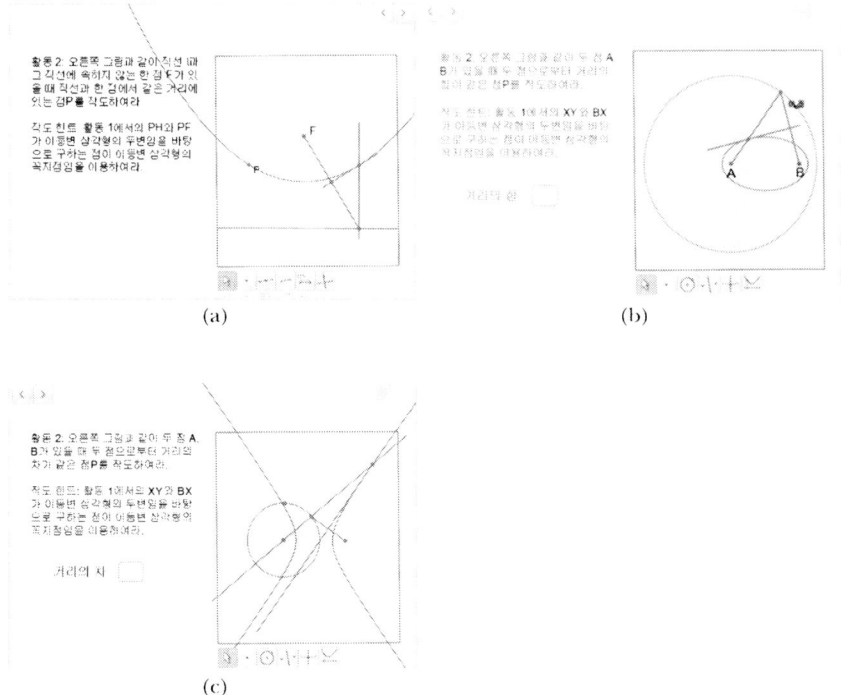

Figure 3.5 Textbook to emphasize interconnection of mathematical representations.

and *PB* is a given value. Figure 3.5c shows the problem to construct a point *P* such that when there exist the given points *A* and *B*, the difference of the lengths of *PA* and *PB* is a given value.

Feature 4: Meaning of Mathematical Concepts

Figure 3.6a shows that if students click the pictures at the bottom of the page, the pictures are enlarged in the right side box to present the parabolic shapes shown in various natural and social phenomena. In this way, students can appreciate the ubiquity of mathematics in their own life. If students click the button at the right side of the bottom of Figure 3.6a, they can watch the 6-minute video-clip made by Korean EBS (Educational Broadcasting Service) to help students appreciate that the parabola, as a mathematical concept, is used in ordinary life. Figure 3.6b shows a cut of the video-clip illustrating that the temperature of a thermometer in a cup of water located at the focus of the parabola increases enough to make it hot water over time.

Figure 3.6 Textbook to emphasize meaning of mathematical concepts.

EXAMINING USE OF THE SMART TEXTBOOK

To examine the use of smart textbook lessons, four 10th grade students were selected (11th grade students had already studied the quadratic curves). Students in the top 10% of their class were invited from one high school near the author's university, Korea National University of Education (KNUE). The students came to the institute to have 3-hour lessons each Saturday afternoon for 3 weeks in 2014. The teacher was a doctoral student involved in the development of the smart book. The trial textbook was uploaded on a Tablet PC for students to use. A LMS was set up on the computer in the classroom for the teacher to monitor all students' activities and their problem solving processes. The LMS allows the teacher and students to share their screens. This LMS provides an opportunity for the teacher to assess students' activities informally by recording them and for students to explore mathematical ideas and communicate with other students and the teacher, which is an important component in the smart textbook environment.

In this smart textbook, three levels of activities were proposed to mark points that are on the same parabola (Figure 3.7). Students have a chance to place finite points (animals) that lie at the same distance from the given point (font) and straight line (river side) in a real space (Figure 3.7a). Based on this activity, students mark points in the Euclidean space (Figure 3.7b) and the Descartes space (Figure 3.7c). The sequence of activities helps students increase their abstract levels for understanding mathematics and their geometric intuition for problem solving.

Students used these three exploring activities to support their explanation that the trace of center points of circles inscribed in the given half circle all fall on part of a parabola. Figure 3.8 shows that one student drew perpendicular lines from a point in the supposed directrix and then drew the perpendicular bisector of the segment connecting the point and the

Developing and Implementing "Smart" Mathematics Textbooks in Korea ▪ **47**

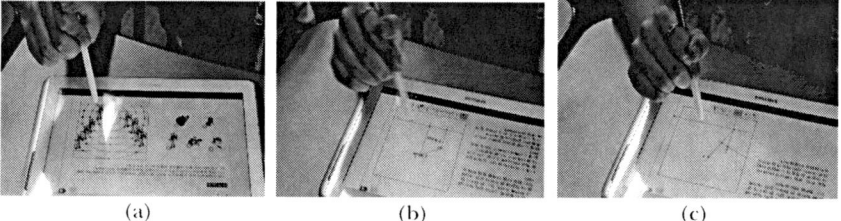

(a) (b) (c)

Figure 3.7 Three levels of activities on real space, Euclidean space, and Descartes space.

center of the half circle to check that the intersection point is the center of the inscribed circle.

To solve a problem on the tablet PC, one student made a guess and drew figures on the digital textbook (Figure 3.8, 3.9a, and 3.9c) but also used the traditional writing method in a paper and pencil environment (Figure 3.9b and 3.9d). Figure 3.9a shows that the student drew the inscribed circle as one example of the problem condition on the digital textbook. Figure 3.9b shows that the student captured the screen to draw a rough figure of a parabola and its focal point and directrix with the pen equipped in the PC. This capturing function of the PC makes its screen a convenient environment for drawing figures like on paper. Figure 3.9c shows that the student tried to find out whether his conjectures met the conditions by drawing on the tablet PC and dragging it. Figure 3.9d shows that the student proved the solution on the figure drawn with the pen.

Another student drew a circle on the tablet PC (Figure 3.10a) that satisfies the problem condition. Then this student used the app of S-NOTE equipped in the tablet PC to represent the relation between the objects, such as the perpendicular bisector (Figure 3.10b). This app helps students represent and record complex mathematical relations on the PC using a pen, which might be difficult to represent using a keyboard of the PC.

Figure 3.8 Problem solving process based on previous activities.

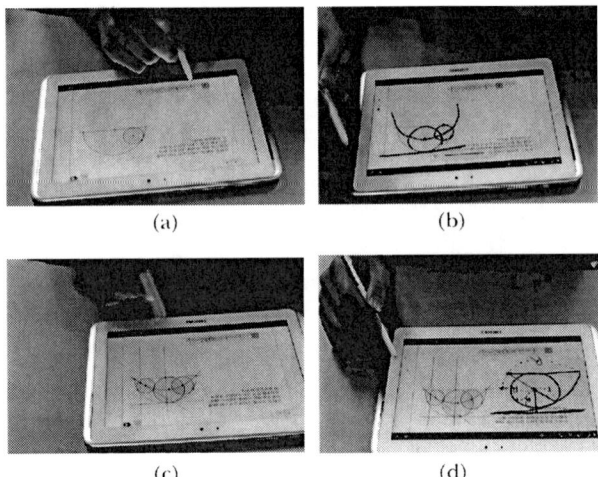

Figure 3.9 Two modes of using digital textbook (1).

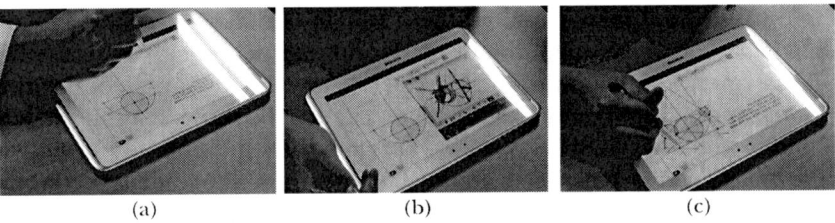

Figure 3.10 Two modes of using digital textbook (2).

Finally, the student drew figures on the PC to confirm whether the relation meets the conditions (Figure 3.10c).

Geometrical activities on the tablet PC support students' solution strategies in a variety of ways. For example, in the problem of finding the focal point when the parabola and its directrix are given, students solved the problem in two ways. One student drew a perpendicular line from one point on the directrix, marked a point of intersection with the parabola, and then drew a circle with the segment connecting the two points as its radius and with the intersection point as its center. The student drew three circles using the same method and determined the intersection point of three circles as the focal point of the parabola (Figure 3.11a).

Students' solution strategies might be due to the activities done before introducing the definition of the parabola. For example, one student drew a line parallel to the directrix and found two points of intersection. Then, by

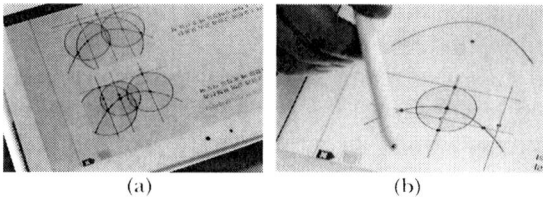

Figure 3.11 Two students' geometrical problem solving.

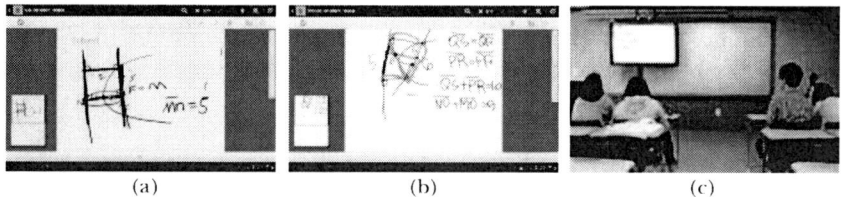

Figure 3.12 Sharing screens for opinion exchange by mirroring.

using the facts that the perpendicular bisector of the line connecting these two points is the axis of the parabola, the intersection of this axis and the parabola is a vertex, and the distance between the focal point and the vertex and the distance between the vertex and the directrix are the same, the student drew a circle to determine the focal point of the parabola (Figure 3.11b). Thus, the student used the symmetry of an arc to answer the question.

A mirroring device was also used to facilitate mathematical communication. Figures 3.12a and 3.12b show students sharing their tablet PC's screen using a mirroring device. Students communicated while they were watching the other students' presentations for solving the problem and accepted or critiqued other students' opinions more easily on the screen (Figure 3.12c).

CHALLENGES

Teachers play a critical role in managing a smart textbook environment. Teachers have to guide students' work, encourage students to communicate with peers, and consider informal assessment based on students' activities, which are all unfamiliar experiences for many teachers in Korea. We need a new kind of teacher training at both the preservice and inservice levels. Action should be a main concept to organize the classroom. To use a smart digital textbook effectively, the facility needs at least 30 tablet PCs and more than 3 cameras, and an LMS system for sharing between students and teacher should be equipped in the classrooms. This is another big challenge. A huge budget is required for these facilities.

Traditional passive teaching and learning practice, in which teachers "talk and show" and students "see and hear," is also a challenge. We have to change teaching practice to be more adaptive to support students' spontaneous activities. However, changing practice is not easy. A digital textbook might serve as the launching point for changing both the teaching and assessment culture in Korean classrooms. To date, only formal written tests have been used for students' assessment. More dynamic tasks operated in digital textbooks need to be developed. Finally, in Korea, whenever curriculum changes, the government issues an evaluation index to be used at the step of textbook approval. The corresponding index for digital textbooks also needs to be developed.

CONCLUSION

This chapter discusses issues, challenges, and benefits in developing and applying a "smart" digital textbook for mathematics education in Korea. The developed smart textbook is an action-based book designed to make mathematics more dynamic and interesting for students. Its fundamental concept is an assimilation process of new knowledge through an interaction between subject and environment rather than through didactic teaching. It is based on Piaget's reflective abstraction. The interaction is facilitated by a self-generating process through direct action, communication among teachers and students, problem solving, and integration of thinking. Furthermore, the smart textbook emphasizes the connection between mathematics and ordinary life, and between geometry and algebra.

However, to manage the smart textbook well, many components of the educational system need to change, including teacher education, educational policy/institutions, assessment culture, curriculum, school facilities, and learning practice. There is no royal road to solve all problems in a single breath. It is our dream to make such an exploring space a reality in today's classrooms—just as Euclid explored mathematics with compasses with his colleagues and students. Since the ancient Greek era, mathematics has been regarded as the embodiment of truth and a tool for finding truth, exploring nature, and developing technology, which has led to a splendid human civilization and culture.

A smart digital textbook should be a math-land in which students can learn mathematics naturally and spontaneously based on reflecting on their own thinking. It helps students manipulate concrete materials with their own perspectives, and develop insight to explore and learn mathematical concepts and solutions through experimental activities. Finally, students come to see the value of mathematics to understand the world around us by recognizing that mathematics is closely connected internally and externally.

ACKNOWLEDGMENT

This work was supported by a 2014 Research Grant funded by Korea National University of Education.

REFERENCES

Drijvers, P. (2012). *Digital technology in mathematics education: Why it works (or doesn't).* Paper presented at the 12th International Congress on Mathematical Education, Seoul, Korea.

Jeong, S. Y. (2014). *Developing T-ACCEPT model as means of support for teachers in a technology-based mathematics classroom* (Unpublished doctoral dissertation). Korea National University of Education.

Lew, H. C., Cho, W. Y., Yu, I. S., Park, W. K., Nam, S. J., Cho, J. M., . . . Jeong, S. Y. (2009). *Geometry and vector.* Seoul, Korea: Mirae & Publishing Company.

Lew, H. C., & Jeong, S. Y. (2014). Key factors for successful integration of technology into the classroom: Textbooks and teachers. *The Research Journal of Mathematics and Technology, 3*(2), 131–150.

Ministry of Education. (1997). *The 7th elementary school curriculum.* Seoul, Korea: Author.

Ministry of Education, Science and Technology. (2011a). *An agenda for action: Smart education as a road to a big nation of talent (Korean).* Seoul, Korea: Author.

Ministry of Education, Science, and Technology. (2011b). *Mathematics curriculum.* Seoul, Korea: Author.

Ministry of Education, Science and Technology. (2012). *A policy on advanced mathematics education.* Seoul, Korea: Author.

Mullis, I. V. S., Martin, M. O., Foy, P., & Arora, A. (2012). *TIMSS 2011 international mathematics report: Findings from IEA's trends in international mathematics and science study at the fourth and eighth grades.* Chestnut Hill, MA: TIMSS & PIRLS International Study Center, Boston College.

Papert, S. (1980). *Mindstorms: Computers, children, and powerful ideas.* New York, NY: Basic Books.

Park, Y. H., Yeo, T. K., Sim, S. A., Kim, S. H., Lee, T. R., & Kim, S. M. (2004). *Mathematics 3.* Seoul, Korea: Chun Jae Publishing Company.

Piaget, J. (1985). *The equilibration of cognitive structures* (T. Brown and K. J. Thampy, trans.). Cambridge, MA: Harvard University Press, (originally published in 1975).

CHAPTER 4

TECHNOLOGY-ENHANCED TEACHING/LEARNING AT A NEW LEVEL WITH DYNAMIC MATHEMATICS AS IMPLEMENTED IN THE NEW CABRI

Jean-Marie Laborde

SOME CABRI HISTORICAL BACKGROUND

Cabri started in 1981 as a research project at Grenoble University. It was the time of first attempts to display images computed on mainframes and displayed on connected CRTs (cathode ray tubes). With Michel Habib (another colleague working in Graph Theory), I created a research project with the aim of realizing a computer-based environment to help us (researchers in Graph Theory) get visual representations of our objects of study, graphs, and computations on graphs.[1]

Especially in research in combinatorics and graph theory, people are accustomed to supporting their thinking with "small sketches" drawn on a sheet of paper, or, for those better organized, on an actual sketchpad (cahier de brouillon in French). So, our project was named "CaBrI" as *Ca*hier de *Br*ouillon *I*nformatique (in English something like Interactive Sketchpad).

One nice example of such a sketch is illustrated here by a *direct manipulation* transformation of the graph in Figure 4.1, constituted of the complete bipartite graph $K_{4,4}$ minus one of its perfect matching. The transformation can change it to a final unexpected shape recognizable as the graph of a cube (or hypercube of dimension 3), as shown in Figure 4.2.

The 1980s was the decade in which more sophisticated graphics software was made possible with the spread of bitmapped screens (Plato IV at CDC (Control Data Corporation), Alto and Star at Xerox, and Lisa and Macintosh at Apple). We had, already, the distinction between "painting programs" and vector drawing programs. On the one hand, "painting" programs (e.g., MacPaint,[2] as shown in Figure 4.3, and later PaintBrush) acted on individual dots arranged to create the illusion of a picture, which was a long-existing technique developed by Romans in their mosaics, used much later by painters such as Seurat (most famously in "Un dimanche après-midi à l'île de la Grande-Jatte"[3]), and then used in the printing industry. On the other hand, there were vector drawing programs (such as MacDraw, as shown in Figure 4.4, and later Adobe Illustrator and Inkscape) acting on a symbolic representation of shapes assembled to create the final picture. Both technologies are valid with their own advantages and, actually, if we look to the current HTML5 standard, both are offered through the support of Canvas and SVG technologies.

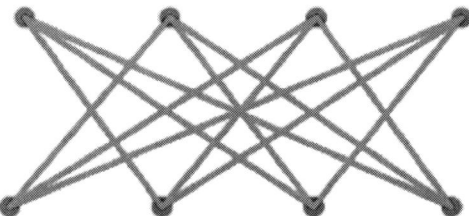

Figure 4.1 A graph.

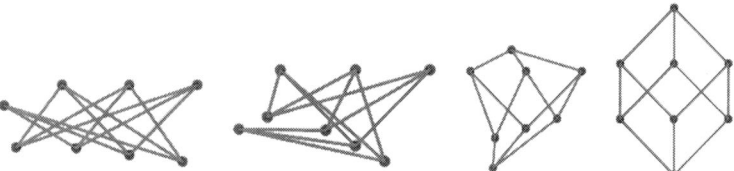

Figure 4.2 The graph in Figure 4.1 transformed into a cube.

Technology-Enhanced Teaching/Learning ▪ 55

Figure 4.3 A Macintosh screenshot with one of the first creations made with MacPaint by artist Susan Kare.

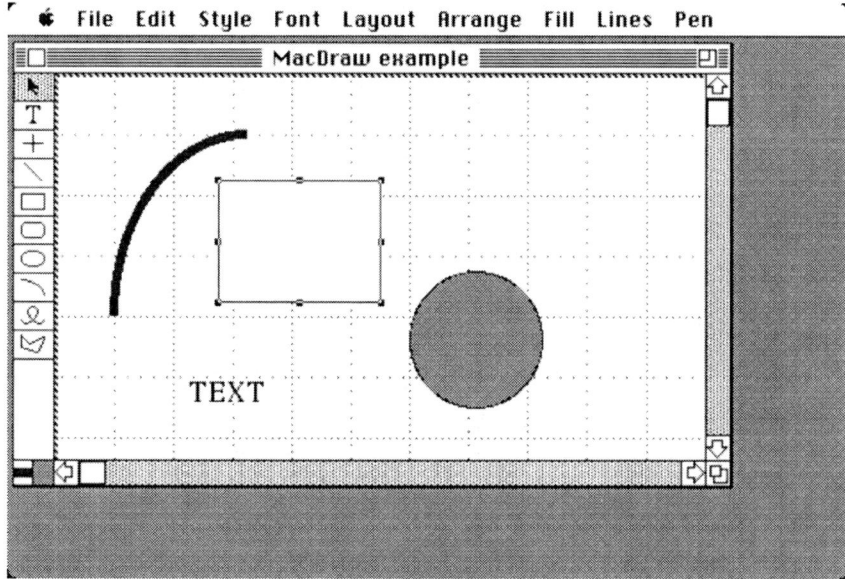

Figure 4.4 An early version of MacDraw, successor of LisaDraw.

Cabri in the early 1980s was designed as a combination of *Vector Graphics* and a *Geometry Engine* that maintained, under the umbrella of the Direct Manipulation philosophy[4] and in real time, the geometric semantic relationships among the various basic components of a geometric figure, say point, lines, segments, and circles. At its birth, Cabri was a "pure microworld" with no explicit educational content, beyond being a microworld supposed to foster learning and enquiry-based teaching as advocated by Papert (1980).[5]

Since then, "presentation software" surfaced rapidly and today is among the software used most often (for better or for worse). Let us mention one of the ancestors, Persuasion (1988), or the widely used PowerPoint (1987), which debuted on Macintosh before being acquired by Microsoft. Such software offers a double interface: one at the *creator* level and another at the *viewer* level. Another creation is HyperCard (see Figure 4.5), combining to a high degree *Direct Manipulation* Graphical User Interfaces (GUI) with *user-centeredness*.

At the end of the 1990s, after digesting various aspects of presentation software (especially of HyperCard), we started to think that Cabri had to change to continue to fulfill, as adequately as possible, what is needed to help people teach and learn science and mathematics. The idea has been to redesign Cabri, this time as a combination of *presentation software*

Figure 4.5 A Hypercard card with its splash screen on top.

(integrating some HyperCard flavor), *microworld philosophy*, and *math engine*, with the whole being governed by innovative *direct manipulation*.

THE NEW CABRI: WHY A NEW CABRI?

The Cabri project started with the emergence of mainframes connected to graphic devices. This gave birth to the GUIs, as opposed to CLUIs (Command Line User Interfaces). The GUI concept was initiated around 1979 with the Xerox Star (Rank Xerox). The Star was rapidly followed by Apple Lisa and Macintosh machines, and later by widely distributed PCs equipped with the Windows interface. In the meantime, technology moved ahead (computing power, screen resolution, number of colors, portability, and so forth).

For instance, on the current iPhone6 screen with its 512×342 pixel resolution, you could put more than 9 original Macintosh screens. If we consider color, we realize that the screen memory of an iPhone is about 3000 times that of the original Mac.

Nevertheless, some educational software from the early age is still in use, like LOGO and, of course, Cabri. Cabri, in forms close to its original, is still used in many countries, including "advanced" ones like Switzerland (Cabri currently has more than 200 million users worldwide).

But going forward, Cabri had to be reborn to take advantage of the latest technology. When speaking about the latest technology, I do not mean the latest technology gadget but instead, I am thinking about real paradigmatic (r)evolutions, like the spread of mobile computing combined, more and more, with tactile GUI on tablets. Those new devices offer handiness, ease of use, and broader affordability.

CABRI FUNDAMENTAL PRINCIPLES

Below are five of the principles that governed the development of the first versions of various Cabri software during the last decades, revisited today in light of progress in learning science and computer technology:

1. *WYSIWYX:* Extending the widely known acronym WYSIWYG (What You See Is What You Get), in the 1990s I proposed the acronym WYSIWYX as a possible refinement. WYSIWYX stands for What You See Is What You eXpect, meaning that the environment should allow the user to perform any kind of action on the objects available on the screen as long it would be reasonably possible to make sense of such specific user action and obtain the result expected by the user. The system would actually perform any user action only as long as it

is compatible with the rules governing the domain. For the time being, of course, no software has managed to implement WYSIWYX to a full extent; it remains a goal that will be reached only after much more research in the cognitive sciences and computer programming has been conducted.
2. *Mathematical fidelity:* The digital environment should in no case display anything contradictory to actual mathematics! This is a special case of "domain fidelity," where we can replace mathematics with any domain of knowledge. So, the behavior of a mathematical software package should, among other things and by default, follow the regular way of writing and expressing mathematics used in textbooks and/or by mathematicians.
3. *Smooth learning curve:* To ease the use of the environment, common habits inherited from everyday practice and acquired in the "real world" should inspire GUI gesture. This can lead to a kind of *augmented reality:* the electronic counterpart of tools used in the real world having possibly more or refined features than their originals.
4. *Learner and novice centered:* The learning environment should not be conceived merely as a "demonstration tool" for the teacher. It is more important to smoothly accommodate the limited mastery of a novice user, while not restricting its use by advanced experts in math, physics, or any domain where *modeling* is key.
5. *Microworld architecture:* The global architecture of the environment has to be rooted in the concept of a microworld that allows the user to perform authentic explorations and constructions. Here the architecture has to meet at interface level with *direct manipulation,* which is a key component in achieving a student's engagement and strong motivation. In the case of the New Cabri, the architecture is actually a two- (or possibly a three-) level microworld architecture.

At the highest level, which we call the *author level,* the user has access to all of the more or less complex features. At that level, an author can design sub-microworlds prepopulated with objects to create an activity book for a student. A student then "plays" the activity at the *student level,* which can be viewed as a content-defined microworld. Cabri then offers a third level, the *teacher level,* where a teacher can vary tasks, designed more generally by the author, and can adapt and customize them according to teaching goals and student needs.

It is important to realize there are not many large-scale studies about the effectiveness of computer technology in education. I know at least of two. The first one (Dynarski et al., 2007) was conducted with funding from the U.S. Department of Education and found no effects for exposing or not exposing children and students to computer-based training. This is not encouraging and might be used (as it has been) to try to dissuade teachers and policymakers from integrating computers into education. A second study originates from the Madrid Urban Community and reports an average performance increase of 25% on the final secondary education exam for a group of 16,000 students using a set of "interactive" software; 402 teachers (not necessarily volunteers) were involved in this experiment lasting six years (Cabezas & Sáez, 2006).[6]

When trying to accommodate such opposite conclusions by two major studies, one soon notices that the pieces of software used in the first study were almost exclusively (if not exclusively) of an expositive software type (i.e., page-turner or other software that we call in France "Math TV,"[7] Math Television, or Math Teaching Video). This type of software does not offer anything fundamentally new; this kind of software started in the 1950s to provide "good" teachers in remote—or less remote—areas through broadcasting.

The four pieces of software involved in the second study were, by nature, different from those of the first study. They shared a common crucial feature: They could all be viewed as specific microworlds. In a microworld type of software, a student can act *directly* on the objects of the domain and, as we will see below, this is an important component of interactivity, especially when this is done under *direct manipulation*.

Direct manipulation takes place when the user can act directly (with the mouse or his fingers) on the electronic counterparts (representations) of objects populating any specific knowledge domain. Acting directly means minimizing any "indirection": for me, changing in web (Java) applet the parameters of the situation (say a or b in a model governed by a quadratic like $ax^2 + bx + c$) through a slider (even acting in real time) is not really *direct manipulation*. Another example stemming from the early 1980s is the way we set the time on a personal computer (or tablet). On a typical Windows machine, you set the time by clicking up or down arrows to move forwards or backwards (see the leftmost image in Figure 4.6). Here, there is at least a three-level indirection: choosing the minute (or second or hour) field is one, accepting the convention that pressing the up arrow means forwards

Figure 4.6 From left to right: Setting time on Windows 7, MacOSX (indirectly), and MacOSX (directly, by moving the needles).

is a second one, and accepting that pressing the down arrow means backwards is a third. On the Macintosh of 1985 (see Figure 4.6), setting the time used to be done by moving the needles directly with the mouse, in direct manipulation (even if strictly speaking manipulating the needles via the movement of a pointing device like the mouse constitutes one level of indirection, disappearing in the case of the tactile interface of most tablets). On the current MacOS (Yosemite) I was surprised to discover something really "à la Windows" (though more elegant) until I discovered that pressing the arrows was a lesser evil (pis aller) and actually that it was also possible to set the time precisely by grabbing the hour-, minute-, or second-hand needles.

INTERACTIVE OR DEEPLY INTERACTIVE?

I have just used the word *interactivity* above, and I was pleased to hear many people using this term when commenting about some of the software presented and discussed during the conference. Many of them were using the adjective *deeply interactive* and not simply *interactive*.

Here I would like to scratch the surface of the widely used and abused term of *interactivity*. *Interactivity* is a very general concept to speak of some of the characteristics of so-called *interactive environments*. The concept of interactivity dates back a very long time. First attempts to clarify probably began with researchers like Norbert Wiener, for whom interactivity was essentially involved when a system was controlled by an action-feedback loop (Wiener, 1948). Wiener stated this explicitly in his book, *The Human Use of Human Beings: Cybernetics and Society*, writing "Thus the theory of control in engineering, whether human or animal or mechanical, is a chapter in the theory of messages" (1954, p. 16). Since then, no satisfactory definition of *interactivity* has reached wide acceptance in the research community, despite an impressive growth of research and literature devoted to the subject. For instance, Thomas Reeves wrote, "I contend that a wiser course would be to support more development research (aimed at making interactive learning work better)... and less empirical research (aimed at determining how interactive learning works)" (Reeves, 1999, p. 19).

Surprisingly, the last decade has seen fewer new publications on the subject; in the meantime, we have been inundated by many computer environments claiming themselves interactive, or highly interactive, in an attempt to attest to their efficiency. So Schulmeister (2003) complained that the Learning Objects Metadata (LOM) project group of IEEE (Institute of Electrical and Electronics Engineers), as well as the ARIADNE (Alliance of Remote Instructional Authoring and Distribution Networks of Europe), consider *interactivity* essentially something to be evaluated on a five-step scale (very-low, low, medium, high, very-high), not telling too much about how to characterize those levels. Then he insisted: "Authors or evaluators of learning programs frequently call such programs "interactive" although the web pages of the learning platform or the screens of their multimedia programs do not in fact contain interactive elements...Interactivity must be strictly distinguished from navigation. To me, interaction means controlling the object, subject, or content of a page" (Schulmeister, 2003, pp. 62, 64, 65).

In this last sentence, we can find part of the essence of the meaning of the *deeply interactive* expression. In controlling knowledge domain objects, the user does not stay at some superficial level of *interactivity*. We have something more than is offered by navigating through pressing nicely decorated buttons, something that does not influence any quality of deep understanding and learning. Deep understanding means not to rest at the surface of the concept, and it goes in parallel with an in-depth interface design, by which I mean an interface that is not thin, but has an actual thickness. Within such an environment, the user, beyond direct manipulation, can really engage in tasks and problems to solve. Conditions to favor this have been described in a book, Kearsley & Schneiderman (1999),[8] where they developed the concept of *direct engagement*. An educational environment being *deeply interactive* is not another way of expressing its *high interactivity*; it is a necessary condition for any in-depth learning and deep understanding. In this respect, the New Cabri presented here and during the conference is aimed (at least according to our wishes) at being *deeply interactive* in bridging our everyday world with the more abstract one that we (learners, students, everybody) build when approaching things mathematically.

A NEW GUI (INTERFACE) FOR THE NEW CABRI

In its first edition, Cabri (1987) used to follow the classic pull-down menus. In 1993, at the time of the famous MacOS System 7, it introduced an entirely new graphical icon system with the top icon matching the actual item from a pull-down menu. The small black triangle at the bottom-right of the icon is intended to be an invitation to press the icon to reveal the full menu (see Figure 4.7). The idea was to avoid cluttering the screen with too many icons and to connect the visual geometric concept embedded in the icon with its verbal representation in words. Later this idea was ported to Cabri

Figure 4.7 Cabri 1 (1993) and Cabri-TI-92 (1995) menu bars.

for the TI-92 and has been used by others including MS-Word (Word 6) and finally Geogebra in 2004. There is no need to mention that this use, in both cases, was made without permission from or giving credit to Cabri.

Ten years later, Cabri II Plus still has the same concept and icon-menu layout (see Figure 4.8). Figure 4.9 shows the Geogebra 4.0 menu bar,[9] which is practically identical to its 2006 version. Figure 4.10 shows the new iconic menu interface for Cabri, with an "express tool bar" for general housekeeping (New, Open, Save, Undo, Redo, ...) sitting on top of the math tool bar, working as before.

SOME OF THE NEW FEATURES OF CABRI

First of all, after years of experimentation, we have come to the conclusion that, because we live and work in three dimensions, it would be more natural to let the students act in a digital environment with a more natural 3D look, even on a flat computer screen or tablet. Basically this is what the New Cabri tries to achieve in allowing students, at any time, to look at their tasks in a more natural way, in 3D with the possibility to switch from pure 2D (with the screen appearing as a sheet of paper) to a 3D representation of the task in a more natural perspective.[10]

In the same orientation, the new Cabri offers, in addition to classic dynamic geometry tools constructing objects more or less magically in one click, tools that are virtual reality electronic counterparts to ordinary physical tools like the compass, ruler, or protractor. See in Figure 4.12 how one can create a circle in

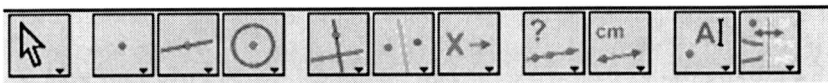

Figure 4.8 Cabri II Plus (2002) menu bar.

Figure 4.9 Geogebra (since 2006) menu bar.

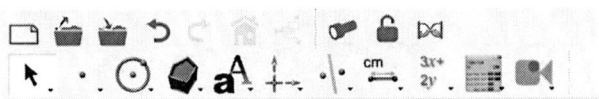

Figure 4.10 New Cabri (since 2010) menu bar.

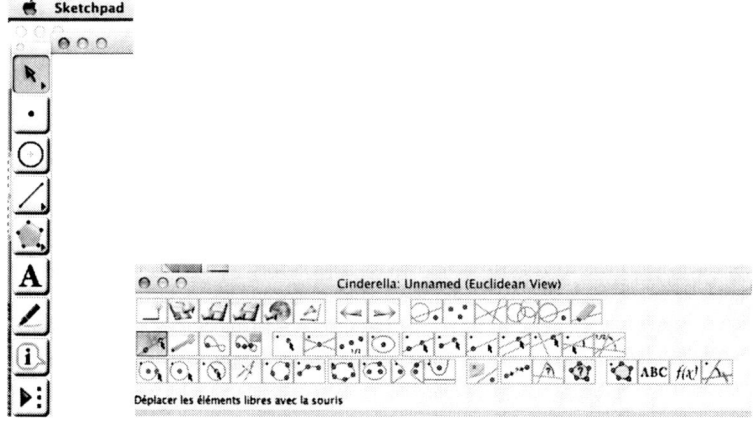

Figure 4.11 Menu bars for Geometer's Sketchpad and Cinderella.

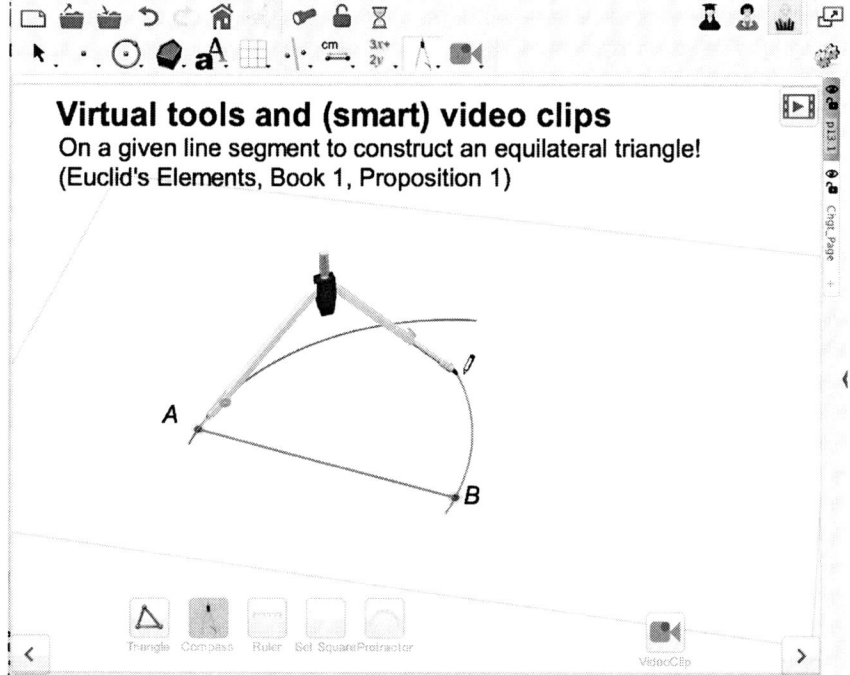

Figure 4.12 Making a circle with the compass tool.

the New Cabri with its virtual compass tool. In Figure 4.12, a virtual compass is used to construct arcs of a circle. Its behavior mimics that of a real compass and the user can change his/her point of view during the drawing, as it would be possible to do on a real sheet of paper. In addition, the compass features some "augmented reality," as its branches will tilt and extend as needed (e.g., to draw a large circle).

A situation where the ability to switch from 2D to 3D that makes especially good sense for students is working with polyhedron nets. Assembling polygons that stick together automatically makes it possible to create any planar net of a polyhedron. A teacher, for instance, can prepare an activity about cube nets. In running the activity, the student will be faced with a slightly "exotic" net to decide if it is the net of a cube (see Figure 4.13). Instead of immediately answering the Yes–No–Don't Know multiple-choice question, the student can experiment by grabbing any face and folding the net in 3D (see Figure 4.14).

Going further in the activity, on the next page the student is asked to create a net of a polyhedron with the form of a kennel using squares and triangles (see Figure 4.15). Figure 4.16 shows the net created by a student, and Figure 4.17 shows the net folded like a kennel.

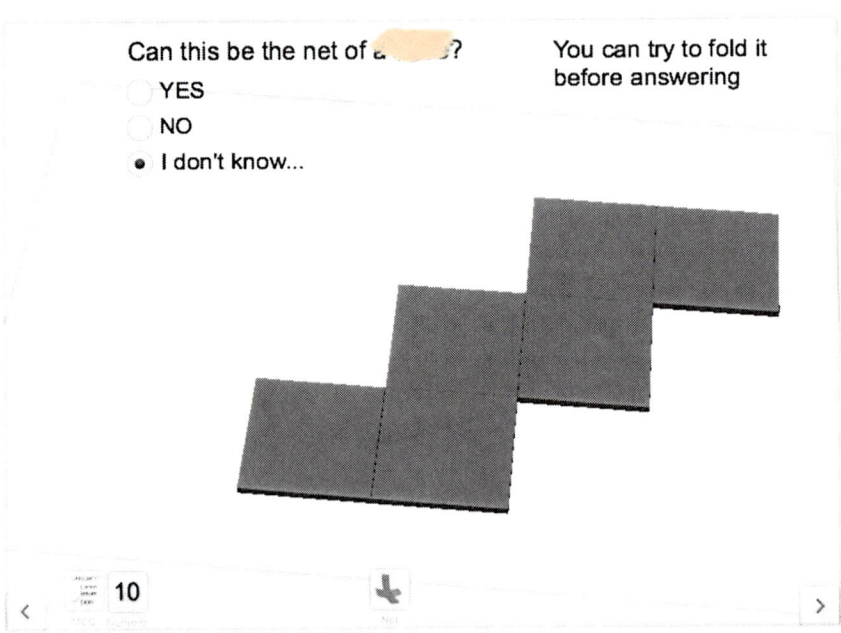

Figure 4.13 A question about a polyhedron net.

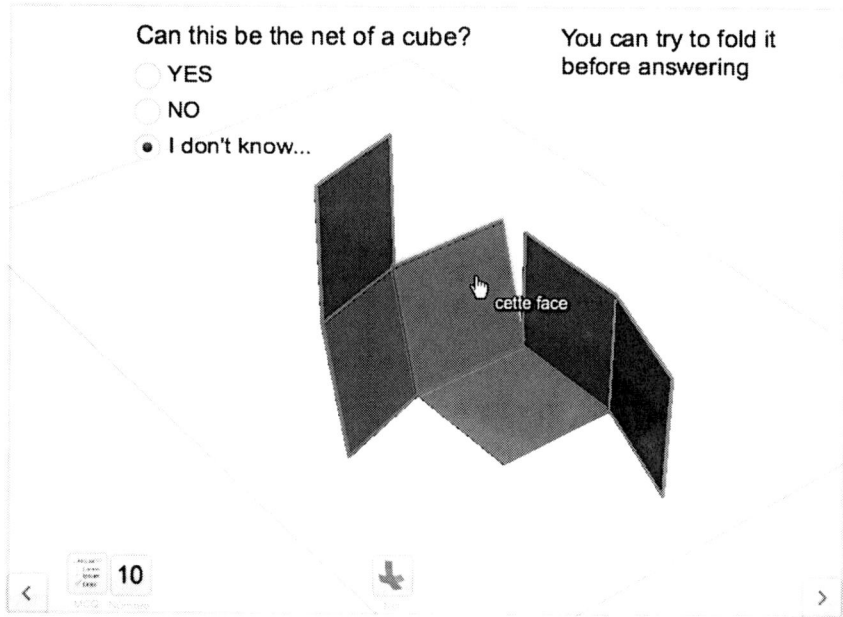

Figure 4.14 Folding a potential net for a cube.

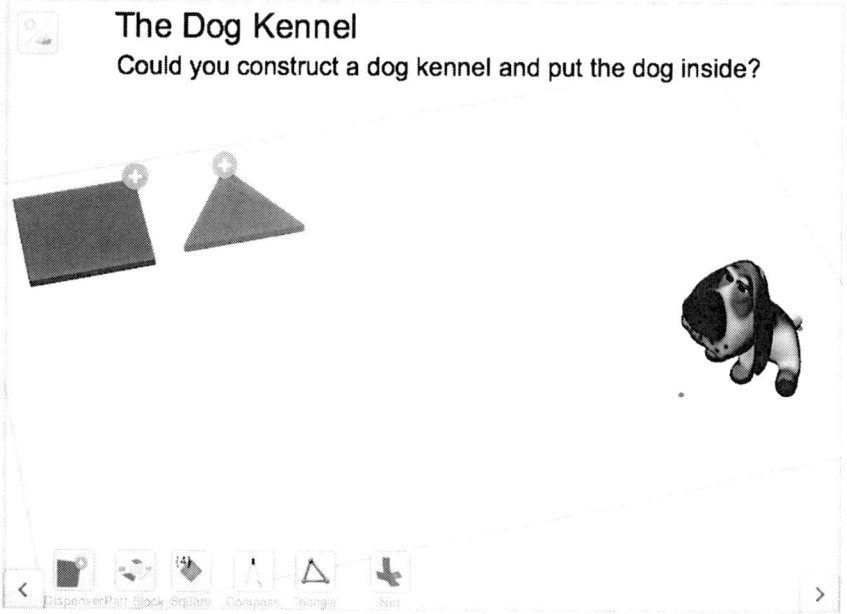

Figure 4.15 A question asking students to make a net in the form of a kennel.

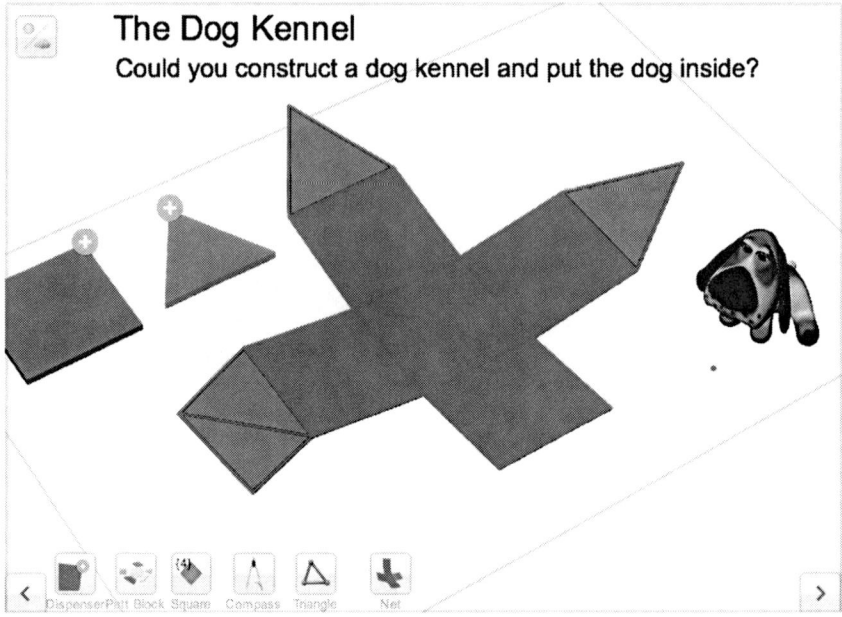

Figure 4.16 A net for a kennel created by a student.

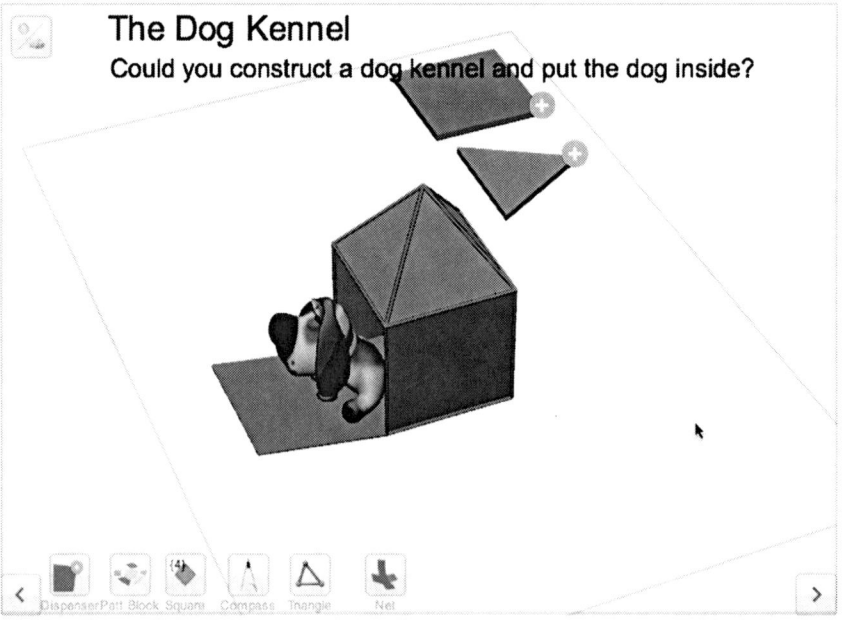

Figure 4.17 The net for a kennel folded.

BRIDGING THE REAL WORLD WITH THE MORE ABSTRACT AND SYMBOLIC WORLD OF MATHEMATICS

First Example

The new Cabri is intended for the teacher to create mathematical activities with an engaging connection to the real world. Because this conference was held along the Midway Plaisance where the World's Columbian Exposition Fair took place in 1893, I would like to develop the following example (see Figure 4.18), where students (perhaps in ninth grade) are asked a "non-traditional" question for which actual mathematical thinking is needed for it to be solved. The question asked is: What is the distance traveled by someone standing in one of the cars of a Ferris wheel during one complete rotation? This raises another question: Does it matter where you stand in the car (in the middle, on the left, or on the right of the car)?

It is a good idea to start with a simplified model of the wheel, say with just 8 cars (see Figure 4.19). At first, a student might realize that when the wheel rotates, the curve described in the air by somebody on the right of the car will interlace with the one described by another person standing on the left. So it might be quite complex to get an idea of the length covered after one turn.

The Midway Plaisance
1893 World's Columbian Fair

What distance will you travel of the Ferris Wheel?

hint Its diameter is 76.2 m
(one foot is 0.304,8 meter)

Figure 4.18 A question about the 1893 World's Fair.

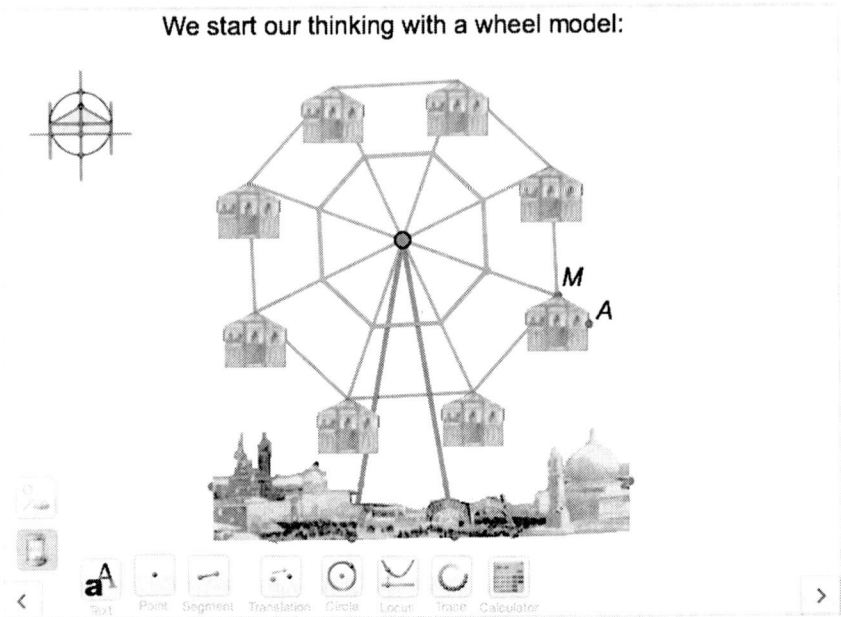

Figure 4.19 A model for solving the problem in Figure 4.18.

Here Cabri can help in offering a trace tool. Let us trace point *A* to represent one passenger (see Figure 4.20). The curve appearing when tracing *A* seems to be a circle, and this is even more clear when students ask for the trace of both *A* and the point *M* where the car is hinged to the whole structure. Point *M* clearly describes a circle, and *A* describes the same curve, just "shifted" from *M* to *A*.[11] Knowing the radius, it is easy to conclude that all passengers, at least if they do not move inside the car, will travel exactly the same distance.

Second Example

Our second example is appropriate to use at any time at the end of secondary education (or possibly earlier). The idea is to apply some of the rules of perspective drawing in a real world context. We assume that students have been previously exposed to these rules, possibly in a setup where the student to some extent discovered them for himself or herself.

The problem for the student is to create a picture of a person visiting a famous landmark, such as the Gyeongbokgung-KeunJeongJeon palace in Seoul (see Figure 4.21).[12] We want to overlay the image on the left with the one on the right. There is clearly a question about the relative size the

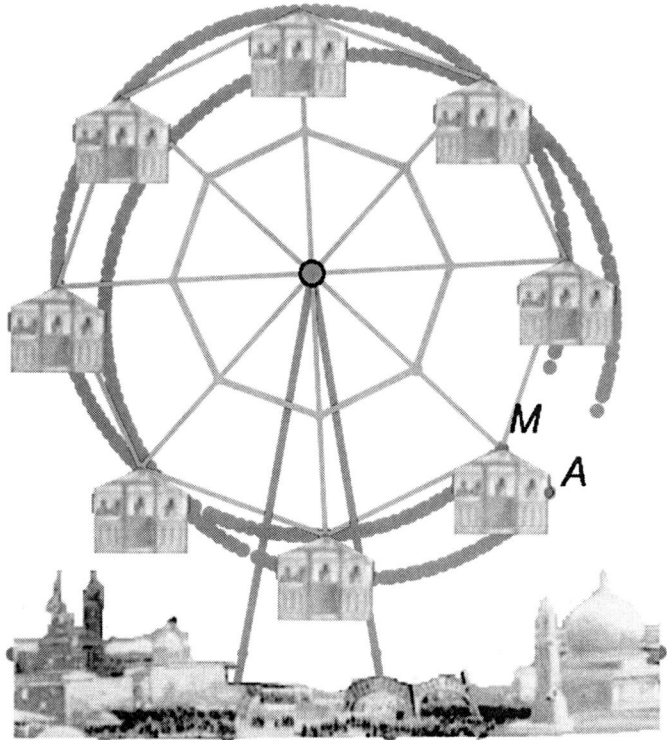

Figure 4.20 A trace of the paths of two riders.

person has to have, when superimposed on the palace background. Too big or too small are equally wrong (see Figure 4.22).

The idea is to make use of some of the perspective drawing rules and to take advantage of the one person already present in the scene close to the stairs. One must use geometry tools in Cabri to "construct" a plausible size for the person when placed inside of the scene.

Among these rules, the following are especially meaningful:

- Vertical lines are represented vertically.
- Front horizontal lines are represented horizontally.
- Parallel lines to any direction are represented as lines converging to some vanishing point.
- The set of all vanishing points of horizontal directions makes a line, the "horizon."

Figure 4.21 The beginning of an exercise in perspective.

Figure 4.22 Figures that are too big and too small in perspective.

Figure 4.23 A solution to a perspective problem.

In Figure 4.23, the student has first re-explored and checked the rules (in green). This has led to a "central" vanishing focal point F. Then, starting at F, she has created two rays passing through the top of the head A and the feet B of the person already in the picture. One can then incorporate the image of the person attached to a vertical segment $A'B'$ with A' anywhere on FA. As a result, one obtains the final placement at a different position and with adequate size $A''B''$ as long as $A'A''$ and $B'B''$ are parallel to the horizon.

CONCLUDING REMARKS

For a long period of time, financial constraints and the objective difficulty of providing "computing power" for every child in a classroom have constituted real barriers towards the adoption of digital technology on a large scale in education. For instance, graphing calculators have been massively adopted because, in most cases, they can be turned off or on instantly, but other technologies have not. Today, this is about to change with the spread of so-called tablets that are easy to carry and can be set up in no time.

A tablet offers a more direct interaction between the user and the digital environment. Quite recently, the mouse represented immense progress in the way people interacted with their devices, as they could manipulate things directly on the screen. But a mouse is an indirect pointing device, and a tablet offers a more direct means of interaction: the finger.[13]

For a long time, running apps on different devices has been a nightmare because of hardware disparity, even if sometimes running apps through the Internet in a browser could be an option. To some extent, the Internet has been a limiting factor for quality in educational environments. The Internet and Java have favored the emergence of many individual apps devoted to some very specialized aspects of math curriculum where user interaction is merely the pressing of buttons and the manipulation of sliders (a slider is typically an indirect means of interaction). HTML5 and associated technologies (Javascript-Canvas-SVG-WebGL) offer painless connectivity when it is needed (i.e., online and offline access) and are very attractive, though not many of the highly and deeply interactive apps can run yet without a plugin.

Nevertheless, due to progress in computer technologies and the cognitive sciences, including in math education, we now gaze at the birth of new, sophisticated educational environments illustrating deep interaction. In short, we have renewed reasons to hope for a brighter and even more exciting education for everyone!

NOTES

1. Here graph means a set of vertices connected with edges, not a function graph.
2. Interestingly *MacPaint* (see Figure 4.3) first originated with the name *SketchPad*, then *LisaSketch,* and *MacSketch* when ported to the Macintosh.
3. Painted in 1884, housed in the Art Institute, Chicago.
4. At least as important as WYSIWIG (see below for definition) is *Direct Manipulation*, as introduced for the Xerox Star Machine, which characterizes Graphical User Interfaces (GUIs) where the user acts on sensible representations of the objects of the domain, rather than acts indirectly by triggering commands to modify the objects. Direct Manipulation has been key in revolutionizing desktop publishing and in the success of environments for teaching/learning like Geometer's Sketchpad and Cabri-géomètre.
5. It is important to pay a tribute to Seymour Papert and remember his invention of this concept of microworld when he created the widely-recognized *Logo*, the archetype of all microworlds. The *Logo* movement launched a whole new way of using computer technology for teaching/learning. If one asks the question why *Logo* is not as popular as it was at its beginning in the 1970s, I believe it is because of its power: being a programming language, *Logo* is quite universal, not connected to any specific teaching "content." So, although it is excellent as a teaching tool in the hands of talented teachers, it unfortunately appears to be too demanding for average teachers.
6. The students were divided into two groups, with and without technology. Technology actually constituted permanent access to four pieces of software,

namely the computer algebra system Derive, the browser Internet Explorer, the spreadsheet Excel, and the dynamic geometry software Cabri. It should be noted that for subject matter closely related to school geometry (as it can be found in the final exam for secondary education), students performed 29.7% better than students from the test group.
7. MathTV is today a commercial brand name registered in the United States.
8. In 1999, Kearsley and Schneiderman established the framework of *Engagement Theory* for technology-based teaching and learning. Beginning their paper, the authors write: "The fundamental idea underlying *engagement theory* is that students must be meaningfully engaged in learning activities through interaction with others and worthwhile tasks. While in principle, such engagement could occur without the use of technology, we believe that technology can facilitate engagement in ways that are difficult to achieve otherwise. So engagement theory is intended to be a conceptual framework for technology-based learning and teaching."
9. It is interesting to note that Geogebra and Cabri II share a surprising resemblance. The other genuine Interactive Geometry Software have clearly distinctive layouts: see Figure 4.11 for Geometer's Sketchpad's global layout and Cinderella's menu bar. Even though they address similar needs from users and offer similar mathematical content, Cabri, Cinderella, and Geometer's Sketchpad have clearly different GUIs. Note that Cabri (like Geometer's Sketchpad) has been the result of practically 20 years of research, whereas the complete Geogebra came out in less than 2 years.
10. We call perspective "natural" when it gives the same cognitive impression as if the objects of the scene were real. For instance, a cube with an edge of 10 cm is represented in central (conic) perspective, as it would be from the eye at a distance of an arm's length. This is the kind of perspective offered by cameras equipped with some standard objective (say 50 mm in the old-time of 35 mm cameras). This is very different from Cavalier Perspective, military perspective, or other exaggerated perspectives created by some extreme field view angle and often confusing for young students.
11. Here we have tried not to use any "advanced" vocabulary, like *locus, translation,* or *vector,* because in some countries, including France, those words are nowadays introduced only with 11th or 12th graders (for better or for worse).
12. The situation is exactly the same as in the movie industry, the director wants to superimpose some characters in a landscape shot previously without them.
13. In some cases, a more precise way of pointing than a finger might be desirable, though Apple was able to demonstrate on iOS and its UI refinement that a finger-based interface can be really efficient. (Multi)finger interaction is clearly more "direct manipulation" than "stylus" interaction. Recall that one of the ancestors of the iPhone as a PDA was the Newton device, driven with a stylus. After its relative failure, Apple decided to base their new UI for iOS devices (iPhones and iPads) on direct finger interaction and not a stylus. To create efficient and natural finger (multitouch) gesture requires a lot of research on hardware, software, and cognition level, and stylus interaction is often a cheap first solution.

REFERENCES

Cabezas, J. M. A., & Sáez, I. M. (2006). Uso de las tecnologías de la información y la comunicación en matemáticas para la ESO y los Bachilleratos. *La Gaceta De La RSME, 9*, 223–243.

Dynarski, M., Agodini, R., Heaviside, S., Novak, T., Carey, N., Campuzano, L., ... Sussex, W. (2007). *Effectiveness of reading and mathematics software products: Findings from the first student cohort.* Washington, DC: U.S. Department of Education, Institute of Education Sciences.

Kearsley, G., & Schneiderman, B. (1999). *Engagement theory: A framework for technology-based learning and teaching.* Originally at http://home.sprynet.com/~gkearsley/engage.htm

Papert, S. (1980). *Mindstorms: Children, computers and powerful ideas.* New York, NY: Basic Books.

Reeves, T. C. (1999). A research agenda for interactive learning in the new millennium. In B. Collis & R. Oliver (Eds.), *Proceedings of EdMedia 99 World Conference on Educational Multimedia, Hypermedia and Telecommunications* (pp. 15–20). Charlottesville, VA: Association for the Advancement of Computing in Education (AACE).

Schulmeister, R. (2003). Taxonomy of multimedia component interactivity, a contribution to the current metadata debate. *Studies in Communication Sciences, 3*(1), 61–80.

Wiener, N. (1948). *Cybernetics, or control and communication in the animal and the machines.* Cambridge, MA: MIT Press. (1961 revised edition)

Wiener, N. (1954). *The human use of human beings: Cybernetics and society.* Boston, MA: Da Capo Press, Houghton Mifflin.

CHAPTER 5

THE RE-SOURCING MOVEMENT IN MATHEMATICS TEACHING

Some European Initiatives

Kenneth Ruthven

THE IDEA OF RE-SOURCING

The term "re-source" was introduced as a means of unfreezing the metaphor of resource: "It is possible to think about resource as the verb *re-source*, to source again or differently. This turn is provocative. The purpose is to draw attention to resources and their use, to question taken-for-granted meanings" (Adler, 2000, p. 207).

In this spirit, "re-sourcing" teaching includes both using conventional resources in new ways and drawing on new sources. The term has become associated with the process through which teachers acquire, transform, and organize material in many forms and from multiple sources to support their teaching, as exemplified by the title of an article in a recent issue of *ZDM*, "Re-sourcing teachers' work and interactions: A collective perspective on resources, their use and transformation" (Pepin, Gueudet, & Trouche, 2013).

Within the teaching profession, there has long been a current of thought which is critical of teaching that is unduly reliant on externally developed schemes of curriculum materials. What might be termed a "re-sourcing movement" has grown up around approaches in which teachers devise their own curriculum scheme through assembling, adapting, and structuring materials from a variety of sources. Major lines of argument for such an approach are as follows. First, by drawing on diverse sources to design new materials or adapt existing ones, mathematics teachers can create a tailored curriculum scheme that both satisfies systemic requirements and takes account of the local context in which they teach. Second, through engaging with the development, evaluation, selection, customization, and organization of resources, teachers can deepen their understanding of how to use them effectively. This type of active and eclectic re-sourcing of mathematics teaching has become increasingly feasible as advances in digital technology have broadened the range of accessible sources and available media, facilitated the preparation and modification of materials, transformed the diffusion and evaluation of open resources, and fostered the creation of interactive user communities.

There has been considerable research on the design, diffusion, and use of conventional resources, particularly textbook-based curriculum schemes. In this chapter, I draw on the much smaller and more recent body of European research on re-sourcing in practice to examine this new phenomenon and issues that it raises. More specifically, reflecting the foci of the research studies that have been undertaken, I examine two aspects of re-sourcing:

- the functioning of online networks for the creation and exchange of mathematics teaching resources; and
- the implementation of a re-sourcing approach by individual teachers, and across an educational system.

THE FUNCTIONING OF ONLINE NETWORKS

A recent study examined French online teacher networks concerned with resource development and exchange in several subject areas, including one mathematics network (Quentin, 2012; Quentin & Bruillard, 2013). This research identified two main prototypes for such networks:

- The *hive*, in which systems of resources are produced by groups of teachers working collectively in an organized manner towards common aims within a carefully regulated framework;

- The *sandbox*, in which discrete resources are produced independently by individual teachers working autonomously with relatively limited regulation.

Sésamath: A Hive Exemplified

Sésamath (http://www.sesamath.net/) is a not-for-profit association in which teachers contribute to the development of systems of open and reusable resources to support lower-secondary mathematics teaching and learning in French schools, including:

- A series of Sésamath textbooks and workbooks, available in downloadable digital or conventional printed form;
- Mathenpoche, an online study-support site for pupils, providing exercises with supporting lessons and animations;
- LaboMEP, an online site that enables teachers to produce customized sequences of resources for their pupils; and
- Sésaprof, an online site for discussion and exchange of resources between teachers.

In late 2012, nearly 20 thousand teachers and over 1 million pupils were registered users of the Sésamath portal, and Sésamath's printed textbooks held 15% of the French market.

Sésamath exemplifies the *hive* type of network. Development of resources takes place through a system of projects guided by the core members of the association. Serving teachers contribute to the association by working together in teams on such projects, according to their expertise. Resources are evaluated and refined through a structured process of peer review and trialing. The resulting resources are attributed to the association rather than to the individuals who created them. The commercialization of printed versions of its textbooks generates income to support the operation of the association. This enables the association to employ staff who provide specialized technical and administrative support (Quentin, 2012; Quentin & Bruillard, 2013).

Indeed, in some respects Sésamath has taken on characteristics of an educational publisher, and this is a topic of debate on its website (see *Sésamath est-il un éditeur scolaire?*). In particular, it has been suggested that there is a "fracture" between the discourses of the "sharing" of resources amongst the contributors to Sésamath and of the "consumption" of its products amongst their users, notably between the commitment to "adaptability" of resources amongst Sésamath contributors against their effort-saving retrieval and use "without any adjustment" by many users. Equally, it has been noted how their participation in Sésamath risks becoming all-consuming for contributors (Quentin, 2012; Quentin & Bruillard, 2013).

Intergeo: More Sandbox than Hive

Intergeo (http://i2geo.net/) arose through an EU-funded project that created an online repository for interactive geometry resources. While the resources themselves were developed by individual members of the wider project community, the project sought to enhance their usability in three particular ways:

- by enabling users to employ their software of choice through specifying a common file format based on open standards;
- by helping users to identify resources that would meet their needs through associating metadata with each resource; and
- by providing prospective users with reviews of each resource through a quality questionnaire completed by previous users (Kortenkamp & Laborde, 2011).

In late 2014, the project had nearly 3000 members and over 30 subgroups, and the repository held nearly 4000 resources and nearly 900 reviews.

Although the introduction of a common file format and of a system for tagging resources with metadata represent *hive*-like tendencies, in other respects Intergeo is more of a *sandbox*. In particular, its repository depends on individual contributions that reflect the personal interests and approaches of the creator rather than arising from any coordinated programme of collective development.

A particular focus of project evaluation was on the functioning of the quality process (Trgalova & Jahn, 2013; Trgalova, Soury-Lavergne, & Jahn, 2011). Data available from the repository showed that only a small proportion of resources (around 13 %) had been reviewed, and that most of these had received only one review. Moreover, few of the resources that had been reviewed had subsequently been modified. Further evidence gathered from relatively active contributors identified a range of reasons why they did not modify a resource:

- They did not find sufficient guidance in a review on how to improve the resource.
- They disagreed with, or were discouraged by, a review.
- They did not feel sufficiently strong membership of the community to want to commit time and effort to revision.

Online Networks From a User Perspective

In both types of networks, teachers emphasized how the sharing of resources between peers inspired them and enabled them to develop and diversify their classroom practice (Quentin, 2012; Quentin & Bruillard, 2013). However, major differences of approach are apparent. On the one hand, Intergeo provides an extensive reservoir of resources and supports users in searching for those that satisfy their own particular criteria, but leaves users to integrate them into a coherent program. On the other hand, Sésamath creates a single complete program, and trials and refines the associated resources to warrant a certain quality, but provides these resources in a form that still allows users to select from and adapt them.

PRACTICING RE-SOURCING AT THE TEACHER LEVEL

Let us turn, now, from the production and exchange of teaching resources through online networks to the implementation of a re-sourcing approach in the immediate practice of teaching. This has been examined in recent research that employs a "documentational approach" to study the organization and use of resources by teachers (Gueudet & Trouche, 2009). To examine re-sourcing at the teacher level, I draw on a recent case study of individual teachers, one (Inga) in Norway and one (Vera) in France (Gueudet, Pepin, & Trouche, 2013).

Inga stated that she usually worked on her own to prepare her lessons, and that there were few opportunities for collaboration with colleagues. She reported drawing on a range of types of resources. First, as well as the textbook used by the class, she consulted other textbooks in the school library in order "to search for ideas." In addition, over her years in the profession, she had collected her "own literature" which provided curriculum-related resources to use, such as ideas for lessons, activities and games, video clips and applets. By adding these to the school's "teaching scheme," Inga could make them known to colleagues who taught similar classes (Gueudet, Pepin, & Trouche, 2013).

Vera regretted the absence of teacher collaboration on lesson planning and preparation of teaching materials that she found in her present school (unlike her previous one). Her "crucial resource" was her teaching folder for each class, in digital and paper form, containing her own plans and copies of exercises/activities from textbooks or other sources. She kept a digital copy of these folders both at home and at school, using a memory stick to transfer between the two. Amongst further personal and school resources that Vera reported drawing on, two textbooks played an important role. The first had been chosen some years ago by colleagues at her school because of its original approach to teaching and learning. The second (the Sésamath textbook) had been chosen with her colleagues the previous year, because of its digital format and special facilities for use with an interactive white board (Gueudet, Pepin, & Trouche, 2013).

Teachers' re-sourcing of their lesson planning appears to rely in the main on an evolving "mashup" of teaching resources. Over their years in the profession, teachers make a strong investment in building up a personal repertoire of resources for lesson planning, drawing on ideas and materials from a range of sources. At the core of these resources lie the teacher's course and lesson plans, linked, in turn, to the teacher's mental "curriculum script" for teaching each topic (Leinhardt, Putnam, Stein, & Baxter, 1991). Nevertheless, textbooks continue to play an important role, providing a default curriculum scheme and a reservoir of relevant teaching material. There is often little collaboration between teachers in the same school on lesson planning, but adopted textbooks act as a common resource, and a school program of work provides a vehicle for sharing other resources.

INSTITUTIONALIZING RE-SOURCING SYSTEMWIDE AT THE SCHOOL LEVEL

These case studies of individual teachers point to a wider organizational dimension, manifested in maintenance of a school program of work, choice of a common textbook, and joint creation of resources. Although this organizational dimension was not strongly developed in these cases, it is often seen as an important one. In England, for example, the collective construction of a common teaching program at the school level has been seen as a major vehicle for supporting the professional development of teachers and the coherence of students' experiences from one year to the next as they move between classes and teachers. In particular, in the English school system, system-wide promotion of this type of organizational strategy has accompanied a sustained attempt to institutionalize a re-sourcing approach to mathematics teaching (Ruthven, 2012, 2013).

The "Scheme of Work" in English Schools

In England, then, schools have been encouraged (since 1989), then mandated (from 2001), to develop their own detailed "scheme of work" for mathematics teaching at each level. As well as serving the internal purposes of providing pedagogical guidance on coverage, sequencing, and treatment of topics and identifying available teaching resources, this scheme of work contributes to the public accountability of the school. In particular, one function of a scheme of work is to demonstrate that the school's teaching of the subject is compliant with statutory national requirements governing curriculum and assessment, monitored through government inspection of the school.

While a scheme of work may draw on a commercial curriculum program, the expectation is that the scheme will show that use of such a program has been locally customized and complemented by a more diverse range of resources. Over recent years, another key factor in the evolution of school schemes of work has been a national drive to equip all classrooms with interactive whiteboard or projection facilities, and the corresponding expectation that schools develop their use of digital resources.

England in International Perspective

Drawing on evidence collected by successive TIMSS surveys, it is possible to put the English case into some form of international perspective. The TIMSS 1995 survey found that secondary mathematics teachers in England were no different from their counterparts around the world in making use of textbooks in their teaching (Beaton et al., 1996). But, by the time of the TIMSS 2011 survey, the English system had become an outlier in this respect (Mullis, Martin, Foy, & Arora, 2012). For purposes of illustration, I compare England to the international mean across the systems participating in TIMMS 2011 and to a select group of individual systems: the United States, Hong Kong, and Singapore.

As Figure 5.1 shows, remarkably few teachers in England use textbooks as a basic teaching resource, and, while the majority continue to use them as a supplementary resource, around 15% report no use of textbooks whatever. Equally, where use of software is concerned (where I imagine that most teacher respondents to the TIMSS questionnaire interpret "software" broadly to mean any type of digital material), England again emerges as an outlier, with over 20% of English teachers reporting using digital material as a basic resource, and virtually all reporting use at least as a supplementary resource (see Figure 5.2). Combining these figures, at most 50% of English teachers report using either textbooks or software as a basic resource, far lower than the international mean or the equivalent figures for the other comparator systems. This confirms the prevalence of some form of re-sourcing approach in English schools.

The English System Under Inspection

England has a national system of school inspection in which schools are visited regularly. In addition, the inspectorate produces periodic

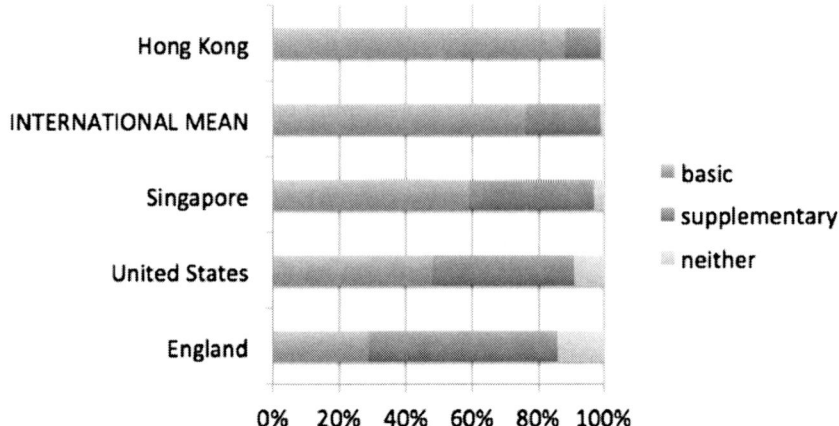

Figure 5.1 International comparison of textbook use: Percentage of Grade 8 students whose mathematics teachers use textbooks as a basic or supplementary teaching resource, or neither. *Source:* TIMSS 2011 (Mullis et al., 2012; Exhibit 8.26).

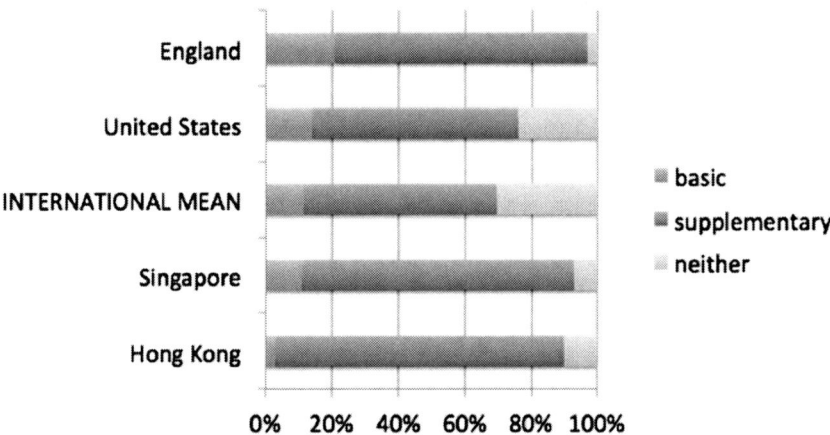

Figure 5.2 International comparison of software use: Percentage of Grade 8 students whose mathematics teachers use software as a basic or supplementary teaching resource, or neither. *Source:* TIMSS 2011 (Mullis et al., 2012; Exhibit 8.26).

state-of-the-system reports. In mathematics, the most recent date from 2008 (OfStEd, 2008) and 2012 (OfStEd, 2012). Although some caution must be exercised in interpreting such reports because contemporary policy preoccupations and imperatives play a part in their framing, they provide probably the best overview available of the system. However, it is important to remember that these documents serve a political function, talking up forms of practice that the inspectorate considered recommendable, and criticizing those viewed as inadequate.

According to the national inspectorate, some schools display many of the qualities attributed to re-sourcing through localized schemes of work: "The best schemes of work included guidance on approaches, interesting activities, and resources that help nurture pupils' understanding. They were seen as living documents, subject to regular discussion and review, which helped staff to develop their expertise" (OfStEd, 2008, p. 25). However, such schools appear to be the exception rather than the rule: both the state-of-system reports contain paragraphs suggesting a minimalist approach to compliance, dependent on externally produced templates: "Good schemes of work were rare. It was not uncommon for teachers to use only examination specifications and textbooks to guide their lesson planning, focusing on content rather than pedagogy" (OfStEd, 2008, p. 25). "Schemes of work were rarely adapted to the particular circumstances of the school and its pupils. They were often simply the schemes provided by [examining] bodies or [by] textbooks" (OfStEd, 2012, p. 73).

The national inspectorate attribute the underdevelopment of school schemes of work to factors such as schools believing their staff do not require such a scheme, or lacking staff with the capacity to develop one: "When the schools had a stable and experienced staff, they frequently did not see the need to formalize guidance... Schools with many inexperienced, nonspecialist and/or temporary teachers, which would most benefit from guidance, lacked the capacity to prepare it" (OfStEd, 2012, pp. 73–74).

Nevertheless, research into the implementation of the reforms that mandated production of a scheme of work indicate that another key factor was the intensification of workload for teachers produced by this and other aspects of the reform: "[Creating] schemes of work continued to be problematic... for mathematics departments, with high degrees of resentment apparent in teachers' statements on the time spent in finding appropriate resources to match objectives" (Barnes, Venkatakrishnan, & Brown, 2003, p. 37).

Extending this line of argument, research on the working life of mathematics teachers has found that two of the strongest predictors of their comfort with the profession and likelihood of remaining in it are satisfaction with "managing [their] workload" and having "freedom to teach in the way [they] choose" (Moor et al., 2006). For many of its teacher advocates, a re-sourcing approach represents a voluntary intensification of their workload in the interests of the exercise of this freedom. However, the shift to mandating the production of a school scheme of work removes the locus of control from the individual teacher; the more that such work takes place under external constraints, the more it is liable to be experienced negatively in terms both of professional freedom and workload intensity.

Indeed, my own experience of a range of schools through my work as a teacher educator and educational researcher suggests that absence of pedagogical consensus within a school mathematics department can be a crucial factor limiting the development and use of a scheme of work. Rather than allowing differences of professional perspective to generate interpersonal conflict and unwelcome constraint, teachers in such a department "agree to disagree" on pedagogy; appealing to respect for professional autonomy, they set a skeleton scheme of work that ensures little more than common pace and coverage, perhaps with further optional references to resources. In this way, the departmental scheme, like the online network, can take on the character more of a sandbox than a hive.

CONCLUDING COMMENTS

The evidence from these European studies shows how—at its best, even if that is not easily achieved—a re-sourcing approach can support the creation, exchange, and use of teaching resources and provide support for the collective professional learning of teachers. However, whether in developmental network or school department, this does depend on propitious conditions, notably the internal generative capacity and shared pedagogical orientation of the collective concerned. At the same time, these studies indicate how a re-sourcing approach risks creating a professional doublebind in which teachers' insistence on developing new schemes of resources prejudices the manageability of their workload—whether in contributing to an online network or developing a local program of work.

From discussions during the CSMC conference, it appears that the informal development and exchange of teaching resources is increasingly prevalent in the United States. In this connection, there was some disparaging talk of the growth of a "Pinterest curriculum," connoting one weak in coherence and driven by superficial choices. On searching online for this phrase, I was intrigued to find that some school districts and educational organizations are already using Pinterest as a mechanism for organizing access to curricular resources! There is scope, then, for a timely research project to investigate the approaches in play and assess their quality.

Indeed, it is rather too easy to imagine that a textbook curriculum is coherent simply by virtue of its design and structure. Rather, what research has highlighted is that textbook use is far more than a matter of translating the sketched-out intentions of the textbook into a full-blown classroom implementation, but involves a process of interpretation, selection, and adaptation by the teacher, so there is no guarantee that the posited depth and coherence of the text are actually realized. Moreover, as the distinction drawn by TIMSS between basic and supplementary resources implies, even where teachers continue to use a textbook as their basic teaching resource, this is supplemented by other resources, increasingly digital, raising the question of how these disparate sources are integrated. This is why we now need a wider-ranging research program on the development, diffusion, and deployment of curricular resources in school mathematics.

REFERENCES

Adler, J. (2000). Conceptualising resources as a theme for teacher education. *Journal of Mathematics Teacher Education, 3*(3), 205–224.

Barnes, A., Venkatakrishnan, H., & Brown, M. (2003). *Strategy or strait-jacket? Teachers' views on the English and mathematics strands of the key stage 3 national strategy.* London, England: King's College/ATL.

Beaton, A. E., Mullis, I. V, Martin, M. O., Gonzalez, E. J., Kelly, D. L., Smith, T. A. (1996). *Mathematics achievement in middle school years: IEA's Third International Mathematics and Science Study (TIMSS).* Boston, MA: Boston College/IEA.

Gueudet, G., Pepin, B., & Trouche, L. (2013). Collective work with resources: An essential dimension for teacher documentation. *ZDM: The International Journal on Mathematics Education, 45*(7), 1003–1016.

Gueudet, G., & Trouche, L. (2009). Towards new documentation systems for mathematics teachers? *Educational Studies in Mathematics, 71*(3), 199–218.

Kortenkamp, U., & Laborde, C. (2011). Interoperable interactive geometry for Europe: An introduction. *ZDM: The International Journal on Mathematics Education, 43*(3), 321–323.

Leinhardt, G., Putnam, R. T., Stein, M. K., & Baxter, J. (1991). Where subject knowledge matters. *Advances in Research on Teaching, 2,* 87–113.

Moor, H., Jones, M., Johnson, F., Martin, K., Cowell, E, & Bojke, C. (2006). *Mathematics and science in secondary schools: The deployment of teachers and support staff to deliver the curriculum.* Slough, England: National Foundation for Educational Research [NFER].

Mullis, I. V. S., Martin, M. O., Foy, P., & Arora, A. (2012). *TIMSS 2011 international results in mathematics.* Boston, MA: Boston College/IEA.

Office for Standards in Education [OfStEd]. (2008). *Mathematics: Understanding the score. Messages from inspection evidence.* London, England: OfStEd.

Office for Standards in Education [OfStEd]. (2012). *Mathematics: Made to measure. Messages from inspection evidence.* London, England: OfStEd.

Pepin, B., Gueudet, G., & Trouche, L. (2013). Re-sourcing teachers' work and interactions: A collective perspective on resources, their use and transformation. *ZDM: The International Journal on Mathematics Education, 45*(7), 929–943.

Quentin, I. (2012). Fonctionnements et trajectoires des réseaux en ligne d'enseignants [Functioning and trajectories of teacher online networks] (Doctoral thesis). Ecole Normale Superieure, Cachan. Retrieved from http://www.stef.ens-cachan.fr/docs/these_quentin_2012.pdf

Quentin, I., & Bruillard, E. (2013, March). *Explaining internal functioning of online teacher networks: Between personal interest and depersonalized collective production, between the sandbox and the hive*. Paper presented at the Annual Meeting of the Society for Information Technology and Teacher Education (SITE 2013). Retrieved from https://docs.google.com/file/d/0BxYfW5Kv7CUqYmEydk53WGJ6UkE/edit?pli=1

Ruthven, K. (2012, April). *Institutionalising the re-sourcing of mathematics teaching: The case of the English school system*. Paper presented at the annual meeting of the American Educational Research Association. Retrieved from http://www.educ.cam.ac.uk/people/staff/ruthven/RuthvenAERA13sigRME.pdf

Ruthven, K. (2013). Métamorphose des ressources et des métiers de l'enseignement: Un bilan anglais. In *cultures numériques, éducation aux médias et à l'information* (pp. 92–95). Paris, France: Centre National de Documentation Pédagogique.

Trgalova, J., & Jahn, A. P. (2013). Quality issue in the design and use of resources by mathematics teachers. *ZDM: The International Journal on Mathematics Education, 45*(7), 973–986.

Trgalova, J., Soury-Lavergne, S., & Jahn, A. P. (2011). Quality assessment process for dynamic geometry resources in Intergeo project. *ZDM: The International Journal on Mathematics Education, 43*(3), 337–351.

CHAPTER 6

INQUIRY CURRICULUM AND E-TEXTBOOKS

Technological Changes That Challenge the Representation of Mathematics Pedagogy[1]

Michal Yerushalmy

THE CHANGING ROLES OF TEXTBOOKS: OBSERVATIONS AND SPECULATIONS

An increasing number of school textbooks are now supplemented by continuously upgraded digital resources that can be found on the Web. Publishers allow teachers to personalize digital textbooks for their courses, emphasizing flexibility and inexpensive dynamic changes, such as selecting from existing chapters and content and even individualizing the book for the student. Hardware manufacturers have encouraged educators to use high-level development tools, and central authorities tout the digital textbook as a unique opportunity for delivering textbooks to distant rural schools. Korea, which has been considered a leading pioneer in the area of e-Textbooks,[2] holds an integrative view in which printed textbooks that are turned into digital format remain the central learning resource, surrounded by other types of facilitating media. Other educational systems are adopting a similar view. But many of the new products are merely digitized versions of their written counterparts and support limited feedback and interactivity, mostly through search and navigation of the digital document. The Israeli education system requires that each textbook be available in at least one of three formats: a digitized textbook, a digitized textbook that is enriched with external links and multimodal materials, or a textbook that is especially designed to work in a digital environment and includes online tools for authoring, learning, and management. Every digital textbook is required to consist of an approved core text that can be appended by personal or external digital resources. A paper textbook is no longer mandatory.

At this time, the most noticeable feature of digital textbooks is the change in the object (delivery format), but, as interactive textbooks of different types appear to be the tools of choice in mathematics instruction in the foreseeable future, it is important to establish criteria for the design and research discourse in regard to student, textbook, and teacher interactions. To illustrate and emphasize central dimensions of e-textbooks and their use by teachers or students, I find it useful to refer to three models of e-textbooks. The models are not distinct but are presented this way for clarity of the discussion, as each marks a key rationale or aspect of functionality.

1. The *interactive e-textbook* is based upon a set of learning objects: tasks and interactives (diagrams and tools) that can be linked and combined; the tasks are based on interactives that are an integral part of the textbook (rather than being "add-on tools"), and the textbook functions only as an e-textbook.
2. The *integrative e-textbook* refers to an "add-on" model where a digital version of a (traditional) textbook is connected to other learning objects that traditionally were not assumed to be part of textbooks, such as learning management, course management, authoring tools to add or edit activities (controlled by teachers), etc.
3. The *evolving e-textbook* refers to a digital textbook that is permanently developing by continual input of practicing teachers or even students (more examples of these types are in Pepin, Gueudet, Yerushalmy, Trouche, & Chazan, 2015).

Each of these three types of e-textbooks can be authored either by an expert author (assuming external authority) or by its users (mainly teachers). Each of the three models of digital textbooks can be either used as is or changed by its users (teachers or students).

In a recent paper, we (Chazan & Yerushalmy, 2014) argue that such technological changes that call for users' involvement pose challenges to the roles played by the textbooks and curriculum materials written by textbook authors and curriculum developers. Following a historical view on the roles and uses of mathematics textbooks, we question whether "mathematics teachers, in addition to textbook authors and curriculum developers, play a more central role in creating curricular materials for students and we expect that though teachers may be able more easily to edit and author documents, they will probably not take the lead in writing textbooks" (p. 64). It is the purpose of this chapter to highlight challenges inherent when teachers attempt to take responsibility for *writing e-textbooks* and to elaborate on the challenges of teachers participating in *editing authored interactive textbooks*. Both modes of participation relate to changes in the traditional authoritarian role of textbooks and the textbook classroom culture.

TEACHERS WRITING E-TEXTBOOKS: SHIFTS OF AUTHORITY

Textbooks are meant to provide guidance to and present opportunities for students to learn by making the objectives and ideas of the curriculum more readily apparent to them. Textbooks also provide guidance to teachers, helping ensure their teaching is in line with the expectations of external authorities. As there is a direct etymological link between "author" and "authority" (Herbel-Eisenmann, 2009; Young, 2007), a textbook's authoritarian position derives from the fact that it is written by a recognized expert or a recognized group of experts.

If we see the textbook as a message about how and which content should be taught, then *coherence is a major requirement.* Worldwide reports draw a common picture: Technological resources that have been populating the web for a long time already—however much they may appear as adjacent to the newer published textbooks—are likely to be seen as enrichment rather than as core. Another documented phenomenon is that successful open-source projects are modular in nature. For example, the Linux open-source programming derives its success from being completely modular, as Bonaccorsi and Rossi (2003) explain. And, as Benkler (2006) distinguishes, online encyclopedias (such as Wikipedia) are modular and can be enriched by the involvement of the crowd while an approved textbook has to adhere to state standards. Although coherence of use is what at the end determines what teachers propose to their students, drawing on the textbook materials, I refer here to the coherence of the design. Designing with open-source repositories would have to reach and proclaim visible coherence, and developing understanding of how coherent a collaborative evolving innovation is may well be the greatest challenge for teachers as authors and users.

The challenge of documenting and evaluating *quality and quality control* increases with the removal of external expert authority. It results in the search for alternatives that can help establish confidence and trust in the resources (Coyne, 2010). Designing ways to evaluate contributions is one of the huge challenges a digital networked culture will have to confront. So far we have only a few examples of how to rethink determining the quality of evolving curricular resources that are contributed by diverse communities. On the assumption that quality is ensured by the participation of large communities and that a crucial part of what makes for high quality open source products is the size of the authoring and participating community, we need to acknowledge that the relatively small educational communities that exist so far have ramifications that challenge acceptable measures of quality. There are not many math textbooks to be found under Wikibooks, for example, and most of these seem to have been authored by a single writer, sometimes with minor contributions from a small group. Other platforms that are offering tools for community writing do not involve large communities of math textbook authors.

Sustainable leadership is another challenge of textbooks authored by communities of teachers. Bonaccorsi and Rossi (2003) state that, "Two factors shape the lifecycle of a successful open source project: a widely accepted leadership setting the project guidelines and driving the decision process, and an effective coordination mechanism among the developers based on shared communication protocols" (p. 1246). Although it seems unlikely that teachers will commit themselves to long-term participation in an evolving project, the example of the French mathematics teachers' online association, known as Sesamath, is an important example. This association (described by Gueudet & Trouche, 2012), dedicated to the design and sharing of teaching resources, was created over a decade ago and has rapidly grown into an online community of practice.

Still another challenge relates to *regulation*. As recently described by Robinson et al. (2014), "New research is needed to understand how the legal permissions associated with open textbooks may enable change in pedagogy, assessment, and student engagements" (p. 350). Setting clear legal norms with regard to open-source materials is part of a wider social movement to create a toolbox to handle the evolution of innovations and products.[3] The unprecedented evolutionary process of authoring that changes the textbook from an object into representing a step in a process reduces the likelihood of a high correlation between the intended and enacted curriculum. Thus, the structural and the normative fundamentals of the new system need to be clear to teachers. In a recent thesis, Har-Carmel (2014) critically analyzes the e-Textbook regulation policy, processes, and norms in Israel and comments on their relevance to other centralized education systems. Har-Carmel explains why three common roles of state regulation of textbooks turn out to be irrelevant: (a) confirmation of consistency and conformity to the standard curriculum, (b) pedagogic and academic evaluation by expert evaluators, and (c) control over the retail process. In e-Textbooks, Har-Carmel explains, the technological evolution of resources reduces the likelihood that consistency and conformity to the state standards will be fulfilled; academic evaluation conducted prior to the granted approval is limited due to the capacity of users to incorporate new resources within the textbook; and, norms and regulation of web materials make the pre-set pricing issue nearly or completely obsolete.

The shift of authority from being externally imposed to residing within the participating community is not unique to textbooks and to education systems, but rather relates to broad social-cultural changes. However, to assume that conditions are set for teachers to take the major lead in writing and producing textbooks and for educational systems to rely upon such contributions seems premature.

TEACHERS EDITING AUTHORED TEXTBOOKS: SUPPORTING TOOLS AND COMPETENCIES

e-Textbook authors often utilize less formal structures that more easily accommodate student-centered guided inquiry. For teachers, adopting books that are authored to deliver curriculum, but are designed with a flexible structure to accommodate inquiry, poses new challenges. One challenge has to do with the extended engagement required of the teacher, who becomes an active designer of the instructional materials. Another involves the capacity of students to learn by engaging in the variety of interactions that multimodal interactive textbooks offer and that were not traditionally part of textbooks. Both are related to the capacity of present-day readers to become involved with complex, multimodal texts and images that require construction of meaning through a synchronization of semiotic and structural resources. Such complex narratives are not rare anymore, and even some traditional printed pages have changed to provide the multimodal feel shared by interactive media readers (Kress, 2003).

An example of a flexible, interactive multimodal digital textbook is implemented in the VisualMath algebra and functions textbook (Yerushalmy, Katriel, & Shternberg, 2002; Yerushalmy, Shternberg, & Katriel, 2014). The design of VisualMath is situated in the larger field of mathematical guided inquiry. The textbook design is characterized by the extensive roles assigned to visual semiotic means, by interactivity between the reader and the mathematical objects and processes, and by a conceptual structure of digital pages that could be rearranged to serve a variety of instructional paths. VisualMath was designed to challenge traditional notions of school mathematics and ways it can be taught and learned. For over two decades, this curriculum has been implemented in a variety of settings in Israel (Yerushalmy, 2013). We envisioned the textbook as a mediator that encourages engagement: of the teacher to plan the instruction in a variety of ways, and of the students challenged to do mathematics supported by multiple representation tools. As the product of academic laboratory development, VisualMath aims to address future innovations and analyze the potential of new technologies and innovative resources. Years before hypertext and digital texts were available, we sought designs that provided incentives for multi-modal and non-sequential reading. The views that led to the design borrowed from the museum setting and were consistent with the distinction Kress and van Leeuwen (1996) made to describe linear and nonlinear texts. They compared linear texts to "an exhibition in which the paintings are hung in long corridors through which the visitors must move, following signs, to eventually end up at the exit," and non-linear texts to an "exhibition in a large room which visitors can traverse any way they like... It will not be random that a particular major sculpture is placed in the center of the room, or that a particular major painting has been hung on the wall opposite the entrance, to be noticed first by all visitors entering the room" (p. 223).

VisualMath was designed through multiple cycles of development and field studies. The findings from these cycles form the basis for the study of three challenging designs of attempting flexibility while enacting digital textbooks: (i) the design of mapping tools for envisioning the terrain, (ii) the use of a visual-semiotic framework for analyzing the affordances of interactive diagrams, and (iii) the use of interactive visual feedback tools for formative assessment. The three aspects grow from the assumption that flexibility (personalization of teaching, design for students, etc.) could occur when (1) the framework of "what is being taught and how to teach it" is clear and its boundaries are well stated: the resources have internal designed coherence, ideas are connected to common cognitive roots, and the resources are designed with pedagogical principles that are well stated; and (2) there are strategies supported by tools to accomplish a relatively simple analysis of the materials and to create, collect, and present data representing students' understanding.

FIRST CHALLENGE: MAPPING TOOLS FOR ENVISIONING THE TERRAIN

The following quotation from Thomas Kuhn (1962) challenges educators in regard to design of curriculum, textbooks, and teaching: "From the beginning of the scientific enterprise, a textbook presentation implies, scientists have striven for particular objectives that are embodied in today's paradigms. One by one, in a process often compared to the addition of bricks to a building, scientists have added another fact, concept, law, or theory to the body of information supplied in the contemporary science text. But that is not the way science develops" (p. 140). It is also not the way constructivist theories envision meaningful learning guided by professional teaching. Brown (2009) compares such teaching to a jazz player's use of the notes, and argues that, rather than designing curriculum materials as one-size-fits-all documents, designers should support different modes of use according to their pedagogical design capacity. Designing mathematics curriculum and the way it is represented in textbooks can be considered an interpretive dynamic process. But does that mean that each teacher would write and rewrite the textbook for the classroom according to different entry points reflected by the students' understanding? How could teachers commit to such a dynamic process? For example, Marty Schnepp (a teacher featured in Chazan, Bethell, & Lehman, 2007) describes why he, as a calculus teacher, had decided to change the order of teaching calculus and specifically reordered the teaching of the product rule and the chain rule. Schnepp was an experienced senior teacher who was part of a group of high-school teachers involved in course-level planning.

Drake and Gamoran-Sherin (2009) argue that teachers gain curricular vision that entails more flexibility when they gain more experience. Vision gained by experience should be fortified and supported by appropriate design. Ball argues that developers should design teacher-sensitive educative materials that support teachers' understanding of trajectories and order while teaching and attending to students' ideas:

> ...designers must have a sensitivity for practice and its demands. For example, in real-time teaching, teachers cannot read detailed instructions as they listen to and interpret learners and manage the trajectory of content through the discourse and activity of the class... Such demands point to the importance of educationally oriented design, based on backward mapping from an understanding of practice and of resources-in-use, or "lived resources," in order to support resource use for improved learning. (2012, p. 351)

For a collection of semi-ordered materials and multimodal digital "pages," which to a certain extent stand on their own, to be considered a textbook, the deep structure of the concepts and the inter-relations between them must be simple and visible. In designing the VisualMath curriculum and e-textbook, we organized the content along a single view of the algebra, focusing on the algebra of functions: VisualMath was designed to use functions as the foundation for mastering algebraic skills with understanding by all students. Organizing the algebra around a single mathematical concept allowed mapping the resources along a relatively small number of mathematical objects and operations. These objects and operations are rich enough to support a variety of progressions and sequences mathematically and pedagogically. However, as my colleagues and I point out in describing our interpretation of functions-based approaches to school algebra, changes of order in presenting content can be far more revolutionizing than often presented (Chazan & Yerushalmy, 2003; Yerushalmy & Shternberg, 2001). We therefore prepared two lists. One consisted of the mathematical objects involved. In its current form, the VisualMath e-textbook accommodates the linear and the quadratic "museum exhibits," each one appearing as a row on the map: a mathematical object. An additional row represents "any" or generic examples of functions. The other list consisted of operations on the functions and with them. The six operations included do not form an exhaustive list. Rather, they are what Schwartz (1995) had called the "interesting middle": operations that represent important mathematical concepts, and are appropriate and useful to learn as part of a function-based school algebra. The operations are *represent* (a function), *modify* (reforming the view or structure of the expression with purpose and without changing the function), *transform* (using operations to transform a function and change it systematically into families of functions), *analyze* the function and its change, *operate* with two functions (synthesizing new functions out of two different or identical functions), and *compare* two functions. The two lists are distinct, and were therefore placed in an orthogonal organization in three layers (as each operation with an object can take place in symbolic, graphic, or numeric representations) of a 2D matrix map, where each cell represents the opportunities for learning resulting from the corresponding operation and object (Figure 6.1). The design considerations are further described in Yerushalmy (2013).

This is not the common structure of mathematics textbooks that usually represent a progression along various themes and views of algebra (Freisen, 2013; Rezat & Straser, 2013). The design of a new organizational map became our major challenge in designing the mathematics and the pedagogy of algebra inquiry; it guided us in posing questions of order, in asking how

Symbolic/ Graphic/ Numeric						
Operation / Objects	Represent	Modify	Transform	Analyze (function change)	Operate (with 2)	Compare
Generic						
Linear						
Quadratic						

Figure 6.1 An organizational map for the VisualMath e-textbook.

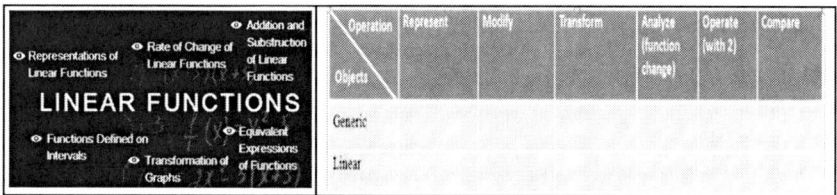

Figure 6.2 The given two-dimensional map for the linear functions' algebra.

known algebra tasks may be based on new resources, and in finding the relevance of traditional tasks that do not seem to be the natural subject to address in an inquiry environment of the type that we envisioned.

The first (2002) version of VisualMath was based on Web 1.0 and had limited participatory means but has been recently redeveloped to be an open interactive textbook (2014). We had asked a few teachers experienced with the VisualMath textbooks to describe how they would attempt to restructure the book to reflect on what and how they would teach linear functions and related algebra concepts. They were given the two-dimensional map for the linear functions' algebra (Figure 6.2) and used it as guidance to organize their teaching: The columns in the personal maps represent the operations that they became familiar with throughout past teaching with VisualMath. The rows in their maps describe their intended curricular sequence (Figure 6.3).

Many of the current sites that offer digital resources also offer indexing according to the Common Core State Standards (e.g., Khan Academy, Better Lessons). Lessons and activities are indexed by the "experience of the crowd" (e.g., tasks are ordered according to the reputation of teachers as task contributors and their teaching experience). I strongly advocate mapping organized by mathematical concepts and structures. An important example of such organization is the work of Confrey, Maloney and Corley (2014), who describe six steps in the creation and evaluation of a learning trajectory that is related to the Common Core State Standards for Mathematics. Further study is required to analyze the appropriate design of flexible textbooks, mapping tools, and professional development programs and materials that would support teachers' understanding of the whole terrain.

SECOND CHALLENGE: ANALYZING PEDAGOGICAL ROLES OF INTERACTIVE TASKS

The issue of constraints embedded in the design of complex interactive tasks was the focus of a long-term research program that Yerushalmy and Naftaliev conducted.[4] A main challenge for the development was to set principles and a framework for designing interactive tasks that, on the one

Teacher #1:
Teaching linear functions and algebra should start with an integrative unit that combines the study of representations of functions and the tools to analyze, define, and represent the ways that any function changes. The linear function is then observed and defined as representing a constant rate of change process. The second integrative unit focuses on the symbolic operations: simplifying and solving equations. Simplifying expressions is viewed as reformulation of equivalent functions and solving equations taught as comparison of two processes. The teacher integrates the study of transformations of functions in this integrative unit in order to introduce the notion of families and parameters. Only towards the end of the topic she suggests to learn in depth operations on equations, arguing that it will serve as the introduction to quadratic functions and equations and the concepts of roots or zeros.

Intended Curriculum or Sequence of Instruction	Represent	Modify	Transform	Analyze	Operate with two	Compare
Linear Function: Representing & Analyzing						
Simplifying & Solving Equations						
Operations & Equations: Preparing for Quadratic						$f(x)\cdot g(x)=0$ $f(x)\cdot g(x)=0$

Teacher #2:
A second plan adopted the function-based view to follow the imperative sequence of algebra curriculum in Israel: Creating a primary basis for multiple representations of processes combined with modification (simplifying expressions). It proceeds to solving equations both algebraically and as comparison of two functional processes. Only later the less traditional approach to algebra is acquired: transformations and parameters, rate of change, operations with functions and the difference function, and the traditional presentation of polynomials.

Intended Curriculum or Sequence of Instruction	Represent	Modify	Transform	Analyze	Operate with two	Compare
Linear Function: Representing & Simplifying						
Solving Equations						
Transformations, operating & analyzing: Preparing for Quadratic						$f(x)\cdot g(x)=0$ $f(x)\cdot g(x)=0$

Teacher #3:
The third plan represents another approach that takes geometric transformations of function's graph to the fore and uses it to view algebraic operations of/on linear expressions as reflecting the geometric transformations. The plan is to get

Figure 6.3 Three cases of intended curriculum structured by experienced VisualMath teachers. *(continued)*

students familiar with ideas of triple representation of function and the idea of slope. Then, it involves study of modifying expressions and solving equations while already having the graphs as a tool to analyze the algebraic objects and get feedback linked to the symbolic actions. Only at the third unit do they study in depth rate of change.

Intended Curriculum or Sequence of Instruction	Represent	Modify	Transform	Analyze	Operate with two	Compare
Linear Function: Representation and Transformations			✓			
Simplifying and Solving Equations						
Operations and Analyzing as Preparation to Quadratic Function						$f(x)\cdot g(x)=0$ $f(x)\cdot g(x)=0$

Figure 6.3 (continued) Three cases of intended curriculum structured by experienced VisualMath teachers.

hand, invited opportunities for active personal learning and, on the other hand, limited changes that pertained either to pedagogy or content (similar to roles of tasks described by Watson & Mason, 2006). We define an *interactive task* as one that is centered around an interactive diagram (ID), which is an applet built around a pre-constructed example. The ID's components are the given example, its representations (verbal, visual, and other), and interactive tools.

The process of active reading encourages readers to continually revise their interpretations, consider alternative perspectives, and even generate and pursue new questions that might go beyond the content of the text itself (Borasi, Siegel, Fonzi, & Smith, 1998). By providing students with opportunities to modify the given examples, to try out their own examples, or to interact with components of the given examples (the representations, as well as the linking and control tools) while reading the text with IDs, negotiation of exploration becomes more explicit because students are encouraged to use various resources to deal with the text. We (Yerushalmy & Naftaliev) focused on the issue of boundaries designed into interactive tasks and found that the design of these boundaries influenced the work of problem solvers and supported students in solving the task.

To analyze roles of interactive tasks in the learning sequence, to design or edit interactive tasks and to analyze students' learning with multimodal resources, we formulated a semiotic framework adopted from social semiotic frameworks of text analysis (Yerushalmy, 2005). We used this semiotic framework to examine the aspects of interactive tasks from the point of view of presentational, orientational, and organizational functions. The organizational function refers to the *connection between all the components of the task*: verbal text, representations, tools, examples, etc. To describe the process of design with IDs, it is useful to look at three types of organization: illustrating, elaborating, and guiding.

The *example that initially appears* in the ID determines the nature of its presentational function. Three types of examples are widely used in IDs: specific, random, and generic. Specific examples that present the exact data of the activity of which they are part provide a dynamic illustration of the text without altering the information. In random examples, a specific example is generated within given constraints, presenting different information at various times and for different users. To serve as a generic example, the diagram must be structured to be representative; it must present a situation that could be part of the given task, but its focus is not on the specific data of the activity.

The *tone in which the text addresses* the learner is subject to design decisions having to do with the orientational function. The "sketchiness" or "rigorousness" of the diagrams are important aspects of reader orientation. An example that appears in a diagram can have an accurate appearance and speak in a strict, distant tone, or it can include a more subtle description and adopt a non-authoritative tone. IDs can function both as sketches and as accurate diagrams. The elements of this framework were valuable in explaining student learning with various interactive diagrams in different contexts of algebra tasks (Naftaliev & Yerushalmy, 2013).

Based on the assumption that the design of the ID establishes the context for a variety of reading, learning, and teaching settings, the major challenge of the study was to analyze how the designed constraints and resources of the IDs function in the process of students developing mathematical knowledge while reading and solving unfamiliar tasks. Three fields in mathematics served as the pivotal points for the design of the problem-solving activities: (a) modeling (analyzing properties of models that mathematize outside mathematical phenomena), (b) formulating mathematical phenomena (writing expressions for linear functions), and (c) manipulating (solving equations). Each objective includes three comparable tasks based on an ID of a different design type and a fourth non-manipulatable comparable diagram: a static diagram or a video clip (Figure 6.4). The activities were assumed to be challenging and new for students and designed to support exploring, conjecturing, and arguing.

The *illustrating* ID was designed to be simple to operate and provided the minimal necessary control for operating the animation. To generate their fuller story, the students looked for ways to bypass and expand the design constraints of the ID. To construct a picture of the motion process, the students resorted to completing the ID using the tools they created themselves, and achieved further understanding with the help of the interviewer. The

Inquiry Curriculum and E-Textbooks ▪ **99**

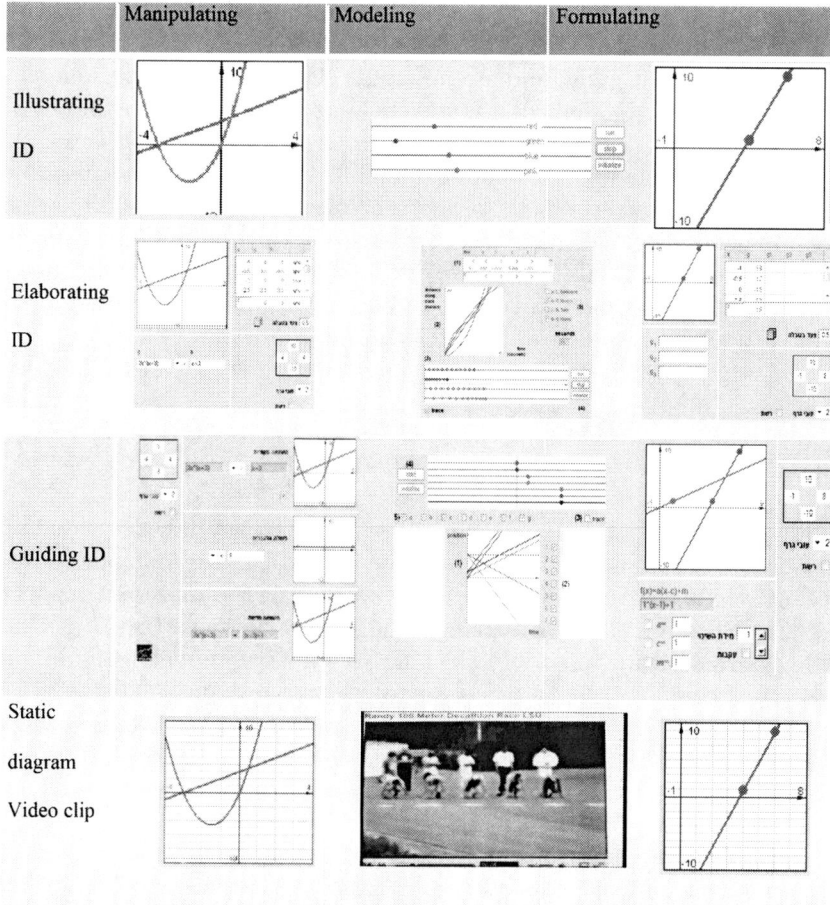

Figure 6.4 The components and design of a comparative study of problem solving with interactive tasks.

guiding ID was designed around a known misunderstanding about a time-position graph describing a motionless object. The students reflected on and modified their current mathematical understandings guided by the designed conflicts and boundaries. Unlike the illustrating setting, which calls for teacher intervention, the guiding ID supports autonomous inquiry. The unusually extensive mental work is noteworthy, as it encompasses an entire cycle of logical argumentation (raising assumptions, deriving conclusions through conjectures and refutations), which led to inventing a fuller story without playing the animation. The variety of linked representations and rich tools in the *elaborating* ID enables various directions of personal choices in viewing the ID as a sketch or as a neat diagram. Naftaliev (2012), summarizing analysis of students' solutions with the elaborating diagrams, argues: "In the elaborating diagram, understanding the generic nature of the animated example advanced with the needs and choices of the students to explore a variety of unfamiliar representations and to interpret the links that were often designed to create uncertainties" (p. 166). The findings accumulated from analysis of students' problem solving processes while using the three types of interactive diagrams in different algebra topics and different pedagogical contexts suggest that this visual-semiotic framework could offer assistance to teachers designing their interactive resources, helping them to identify the type of available resource and adjust its use appropriately.

THIRD CHALLENGE: FORMATIVE E-ASSESSMENT TOOLS AND REPRESENTATIONS

Responsible revisions and adjustments of materials can only be carried out based on experience and close analysis of the learning outcomes. Textbooks' regulation based on external authority claims to represent expertise in the field and is backed by cycles of testing, evaluating, and revising. Once the external authority is altered by internal editing and personalization for the class, teaching can only be assumed to be a responsible action if it is accompanied by continuous evaluation. Thus, continuous formative assessment that informs teachers in various settings about the effects of learning an edited unit is a basic necessity. Addressing this challenge in a curriculum that opts for learning through guided inquiry is even harder, as even experienced teachers repeatedly describe situations of "losing control" (discussed in Yerushalmy & Elikan, 2010). For different reasons, they feel that they don't have enough knowledge about their students' knowledge in the complex environment that includes both individual and small group interactions. Although new technologies give rise to learning new skills, such as problem solving, creativity, critical thinking, and risk-taking, e-assessment has focused primarily on testing factual knowledge. Scalise and Gifford (2006) found that first-generation e-assessment was limited to multiple choice questions, subsequently enhanced by short verbal or numeric answers, and did not reflect the skills and the content and tool interactions that were taught. The need for e-assessment of complex processes, such as conjectures and logical argumentation, and the importance of understanding the role of examples and the links found between interactions with dynamic figures and logical argumentation, are primary motives for the design of new e-assessment instruments.

Our goal is to use technological capacities to support teachers to routinely identify central patterns of knowledge in classes where the curriculum is constantly offering technology-based inquiry. The VisualMath Assessment Toolbox project at the University of Haifa (http://assess.gigaclass.com) is implementing findings in regard to problem solving processes exhibited with the three formats of interactive tasks. We attempt to design a formative assessment environment suitable for supporting various representations of immediate feedback to teachers teaching with the VisualMath philosophy. The assessment environment is built upon the following:

- items that require action with interactive resources (relatively simple guiding or illustrating interactive diagrams, such as dynamic geometry constructions and functions' graphing) that support work in multiple linked representations (constructions, measurements, tables of values); each item takes advantage of the resources both in the challenge that cannot be communicated on paper and the expected actions that provide interactivity and reflect mirror feedback of the actions;
- answers given throughout dynamic interaction with drawings, figures, graphs (of geometric constructions or graphs) to exhibit reasoning processes (mainly by construction of supporting or contradicting examples and submission of conjectures by mathematical representations);
- answers assessed (evaluated and analyzed) automatically; and
- personal and group feedback presented by different representations indicating preconfigured categorized qualities and misconceptions.

Below is an example that we studied in secondary school geometry classrooms. It is of one of the items assessing reasoning, argumentation and proving in geometry (Figure 6.5). It requires interaction with a dynamic figure, viewing shapes and measurements, and presenting the "best examples" to either support or contradict the given statement.

A group feedback sheet that is designed to display the collective answers and results of 28 students for a single item is one of the feedbacks' representations. The DGE interactive item requests a supporting example for two congruent triangles or to argue that there is no such example (Figure 6.6). The automatic checking identifies expected pre-configured misconceptions captured and marked (in yellow) and mistakes (in red).

The visual comparative feedback helped us quickly identify central patterns of knowledge in class. In a discussion that took place in class after the examination, we displayed the group feedback sheet and found that it encouraged active discussion, as students were eager to participate in a conversation based on their generated examples. Successful brainstorming was conducted on the existence of multiple correct answers, the importance of a large-scale example space, and the use of an active and fearless

In the diagram ABC is an isosceles triangle (AB=AC)
DA, DB and DC are the bisectors of angles A, B and C correspondently
1 Gil conjectures that the bisectors always divide the triangle into three congruent trianiges
   Generate a counter example to Gil's conjecture and click Submit
   or click none if there is no counter example
2 Shai conjectures that one of the triangles created by the bisectors can be an acute triangle
   Generate a supporting example to Shai's conjecture and click Submit
   or click none if there is no supporting example

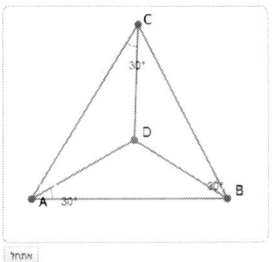

Figure 6.5 Reasoning with examples tasks (Luz & Yerushalmy, 2015).

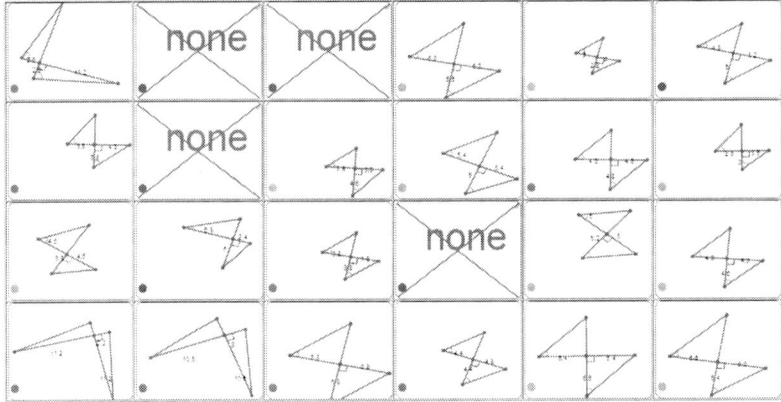

Figure 6.6 Group feedback display (adapted from Luz & Yerushalmy, 2015).

dragging strategy for expanding it, the role of extreme examples, and the need for students to justify a "no such example" answer. The automatic checking also provided quantitative performance analysis and data addressing personal students, and allowed the teacher to take immediate actions responding to students' needs. Other formats of feedback can be used for peer assessment in small groups (online virtual groups or in class).

CLOSING NOTE: THE NEAR FUTURE OF E-TEXTBOOKS AS CARRIERS OF INQUIRY CURRICULUM

Teaching with an interactive textbook should be considered more than a technological change; indeed, it is an attempt to create new paths for the construction of mathematical meaning. An important question to explore is whether teachers' tasks "to critically analyze curriculum materials; to examine the mathematical and pedagogical assumptions implicit in their design;

and to consider how curriculum materials might be read, used, and adapted" (Remillard & Bryans, 2004, p. 386) will take on a different dimension in the new era of interactive textbooks and digital learning resources. Technological advancements, mainly of Web 2.0 participatory tools and norms that seem to be consistent with constructivist pedagogies, pose a challenge to the accepted function of the textbook as a message completed in the past by a recognized external authority that must be unpacked by learners and teachers. Teachers everywhere create teaching materials, search through professional materials available on the Web, and share their repositories and ideas through on-line social networks. It is important to develop and offer tools that support teachers in selecting their teaching materials and designing opportunities for different students or for different didactical purposes (e.g., initiating ideas leading to non-familiar mathematical concepts, practicing activities, or designing assessment). Mapping and navigating tools for understanding the complex terrain are needed. In this chapter I have reviewed three critical tools appropriate to the inquiry philosophy: tools for mapping learning opportunities and trajectories along mathematical objects and operations, tools for analyzing the pedagogical affordances of interactive tasks, and tools that support ongoing assessment. This area of study is becoming increasingly important, as the feasibility of using interactive textbooks, e-assessment formats, and other live resources is on the rise, and with increasing opportunities for teachers to act as designers and authors of interactive materials.

NOTES

1. This research was supported by the I-CORE Program of the Planning and Budgeting Committee and The Israel Science Foundation (1716/12).
2. See http://blogs.worldbank.org/edutech/korea-digital-textbooks. But also see Lew (2016) for difficulties regarding digital mathematics textbooks in Korea.
3. "...the OPEN in open educational resources refers to the fact that these educational materials are copyright licenses that allow anyone to freely 're-use, revise, remix, and redistribute'..." (Robinson, 2014, citing Hilton, Wiley, Stein, & Johnson's (2010) work).
4. See https://sites.google.com/site/interactivediagrams/.

REFERENCES

Ball, D. L. (2012). Afterwords: Using and designing resources for practice. In G. Gueudet, B. Pepin, & L. Trouche (Eds.), *From text to "lived" resources: Mathematics curriculum materials and teacher development* (pp. 349–352). New York, NY: Springer.

Benkler, Y. (2006). *The wealth of networks: How social production transforms markets and freedom.* New Haven, CT: Yale University Press.

Bonaccorsi, A., & Rossi, C. (2003). Why open source software can succeed. *Research Policy, 32,* 1243–1258.

Borasi, R., Siegel, M., Fonzi, J., & Smith, C. F. (1998). Using transactional reading strategies to support sense-making and discussion in mathematics classrooms: An exploratory study. *Journal for Research in Mathematics Education, 29*(3), 275–305.

Brown, M. (2009). The teacher–tool relationship: Theorizing the design and use of curriculum materials. In J. Remillard, B. Herbel-Eisenmann, & G. Lloyd (Eds.), *Mathematics teachers at work: Connecting curriculum materials and classroom instruction* (pp. 17–36). New York, NY: Routledge.

Chazan, D., Bethell, S., & Lehman, M. (Eds.) (2007). *Embracing reason: Egalitarian ideals and high school mathematics teaching.* New York, NY: Taylor Francis.

Chazan, D., & Yerushalmy, M. (2003). On appreciating the cognitive complexity of school algebra: Research on algebra learning and directions of curricular change. In J. Kilpatrick, D. Schifter, & G. Martin (Eds.), *A research companion to the principles and standards for school mathematics* (pp. 123–135). Reston, VA: National Council of Teachers of Mathematics.

Chazan, D., & Yerushalmy, M. (2014). The future of mathematics textbooks: Ramifications of technological change. In M. Stochetti (Ed.), *Media and education in the digital age: A critical introduction* (pp. 63–76). New York, NY: Peter Lang.

Confrey, J., Maloney, A. P., & Corley, A. K. (2014). Learning trajectories: a framework for connecting standards with curriculum. *ZDM, 46,* 719–733.

Coyne, R. (2010). *The tuning of place.* Cambridge, MA: The MIT Press.

Drake, C., & Gamoran-Sherin, M. (2009). Developing curriculum vision and trust: Changes in teachers' curriculum strategies. In J. Remillard, B. Herbel-Eisenmann, & G. Lloyd (Eds.), *Mathematics teachers at work: Connecting curriculum materials and classroom instruction* (pp. 121–337). New York, NY: Routledge.

Friesen, N. (2013). The past and likely future of an educational form: A textbook case. *Educational Researcher, 42*(9), 498–508.

Gueudet, G., & Trouche, L. (2012). Teachers' work with resources: Documentational geneses and professional geneses. In G. Gueudet, B. Pepin, & L. Trouche (Eds.), *From text to "ived" resources: Mathematics curriculum materials and teacher development* (pp. 23–42). New York, NY: Springer.

Har-Carmel, Y. (2014). *Critical analysis of e-textbook regulation policy* (Unpublished master's thesis). The University of Haifa, Israel.

Herbel-Eisenmann, B. A. (2009). Negotiating the "presence of the text": How might teachers' language choices influence the positioning of the textbook? In J. T. Remillard, B. A. Herbel-Eisenmann, & G. Lloyd (Eds.), *Mathematics teachers at work: Connecting curriculum materials and classroom instruction* (pp. 134–151). New York, NY: Routledge.

Hilton, J., Wiley, D., Stein, J., & Johnson, A. (2010). The four "R's" of openness and ALMS analysis: Frameworks for open educational resources. *Open Learning, 25*(1), 37–44.

Kress, G. (2003). *Literacy in the new media age.* London, England: Routledge.

Kress, G., & van Leeuwen, T. (1996). *Reading images: The grammar of visual design.* London, England: Routledge.

Kuhn, T. (1962). *The structure of scientific revolutions.* Chicago, IL: University of Chicago Press.

Lew, H. (2016). Developing and implementing "smart" mathematics textbooks in Korea: Issues and challenges. In M. Bates & Z. Usiskin (Eds.), *Digital curricula in school mathematics* (pp. 35–51). Charlotte, NC: Information Age.

Luz, Y., & Yerushalmy, M. (2015). *E-assessment of geometrical proofs using interactive diagrams.* In the Proceedings of CERME (Congress of European Research in Mathematics Education) 9. Prague, Czech Republic.

Naftaliev, E. (2012). *Interactive diagrams: Mathematical engagements with interactive text* (Unpublished doctoral dissertation). The University of Haifa, Israel.

Naftaliev, E., & Yerushalmy M. (2013). Guiding explorations: Design principles and functions of interactive diagrams. *Computers in the Schools, 30*(1–2), 61–75.

Pepin, B., Gueudet, G., Yerushalmy, M., Trouche, T., & Chazan, D. (2015). E-textbooks in/for teaching and learning mathematics: A potentially transformative educational technology. In L. English, & D. Kirshner (Eds.), *Handbook of international research in mathematics education (3rd ed.)* (pp. 636–661). New York, NY: Routledge.

Remillard, J. T., & Bryans, M. (2004). Teachers' orientations toward mathematics curriculum materials: Implications for teacher learning. *Journal for Research in Mathematics Education, 35*(5), 352–388.

Rezat, S., & Straser, R. (2013). Mathematics textbooks and how they are used. In P. Andrews & T. Rowland (Eds.), *Masterclass in mathematics education: International perspectives on teaching and learning* (pp. 51–62). London, England: Bloomsbury.

Robinson, T. J., Fischer, L., Wiley, D., & Hilton, J. III (2014). The impact of open textbooks on secondary science learning outcomes. *Educational Researcher, 43*, 341–351.

Scalise, K., & Gifford, B. (2006). Computer-based assessment in e-learning: A framework for constructing "intermediate constraint" questions and tasks for technology platforms. *The Journal of Technology, Learning, and Assessment, 4*(6), 45.

Schwartz, J. L. (1995). The right size byte: Reflections of an educational software designer. In D. Perkins, J. Schwartz, M. West, & S. Wiske (Eds.), *Software goes to school* (pp. 172–182). New York, NY: Oxford University Press.

Watson, A., & Mason, J. (2006). Seeing an exercise as a single mathematical object: Using variation to structure sense-making. *Mathematical Thinking and Learning, 8*(2), 91–111.

Yerushalmy, M. (2005). Functions of interactive visual representations in interactive mathematical textbooks. *International Journal of Computers for Mathematical Learning, 10*(3), 217–249.

Yerushalmy, M. (2013). Designing for inquiry in school mathematics. *Educational Designer, 2*(6). Retrieved from: http://www.educationaldesigner.org/ed/volume2/issue6/article22/

Yerushalmy, M., & Elikan, S. (2010). Exploring reform ideas of teaching algebra: Analysis of videotaped episodes and of conversations about them. In R. Leikin & R. Zazkis (Eds.), *Learning through teaching mathematics: Development of teachers' knowledge and expertise in practice* (pp. 191–208). Dordrecht, Netherlands: Springer.

Yerushalmy, M., Katriel, H., & Shternberg, B. (2002). *The functions' web book interactive mathematics text*. Israel: CET–The Centre of Educational Technology. (Available from http://www.cet.ac.il/math/function/english (with *Explorer* only).)

Yerushalmy, M., & Shternberg, B. (2001). A visual course to functions. In A. Cuoco & F. Curcio (Eds.), *The roles of representations in school mathematics. The 2001 Yearbook of the National Council of Teachers of Mathematics* (pp. 251–268). Reston, VA: National Council of Teachers of Mathematics.

Yerushalmy, M., Shternberg, B., & Katriel, H. (2014). *The VisualMath functions and algebra e-textbook*. Retrieved from http://visualmath.haifa.ac.il/

Young, S. (2007). *The book is dead: Long live the book*. Sydney, Australia: University of New South Wales Press.

PART II
IMPLEMENTING DIGITAL CURRICULUM

CHAPTER 7

CONNECTIONS AND DISTINCTIONS AMONG TODAY'S DIGITAL INNOVATIONS AND YESTERDAY'S INNOVATIVE CURRICULA

Valerie L. Mills

The Common Core State Standards for Mathematics (CCSSM; National Governors Association Center for Best Practices & Council of Chief State School Officers, 2010) and digital innovations in mathematics education are beginning to drive significant and widespread changes in curriculum and teaching practices across the K–12 landscape in the United States. As curriculum designers and educators who have already begun to grapple with these changing conditions, we are reminded of a time 30 years ago when mathematics education was similarly working toward broad curricular and instructional changes with an equivalent intensity.

The decade of the 1980s saw the release of numerous reports from the National Research Council, the Conference Board of the Mathematical Sciences, and a variety of assorted commissions. These reports clearly articulated the changing needs of our society, including the need for all Americans to have a strong mathematics foundation, and the nature of the mathematics they would need to matriculate into the world of work. Based on these needs and a synthesis of research findings, the reports called for a series of immediate changes to be made in K–12 mathematics curricula across the nation. This decade of dialogue culminated with the release of the *Curriculum and Evaluation Standards for School Mathematics* by the National Council of Teachers of Mathematics (NCTM, 1989).

The flurry of reports and recommendations in the 1980s prompted the development and adoption over the next decade of what came to be called *innovative* or *Standards-based* instructional resource materials. These resources included collections of more complex problem-solving tasks that were used to supplement instruction and newly design-engineered textbooks and assessment materials. School systems, leaders, teachers, students, and families responded to the implementation of these innovations and for two decades now, as a community, we have lived and learned about the work needed to produce significant change in education.

Today in the United States, we are poised on the cusp of another wave of educational innovation. Again, we are driven by the widespread adoption of new standards, this time the CCSSM, and by the development of newly aligned instructional resources, most of which are infused with technology. Given the similarities with earlier reform efforts, it seems productive to reflect upon some of the lessons learned implementing these changes a generation ago so that we might consider similarities and distinctions between the two environments and then consider if the comparisons suggest possible recommendations for today's work.

The comparisons that follow are taken from the point of view of a practitioner who has been responsible for using innovative curricula, contributing to the development of these new curricula, and overseeing the selection and implementation of instructional materials in schools. The two areas chosen for exploration were selected for their relevance to both practice and the CSMC conference's audience of curriculum developers. They include first, the curriculum resource selection process and second, the work of implementing new curriculum resources. Each section explores lessons learned from previous efforts, including a comparison of similarities and distinctions to today's educational environments, and offers a few related recommendations.

COMPARISON I: CURRICULUM RESOURCE SELECTION

Lessons Learned

Practitioners in the field learned very quickly that the nature of the 1989 NCTM Standards required parallel changes in the process by which textbooks were evaluated and selected (i.e., traditional review and pilots simply were not sufficient and sometimes counterproductive given the nature of the new materials). The change was needed because analyzing the degree to which instructional resource materials (textbooks) were aligned to the new standards could no longer be accomplished by a simple content-based checklist (present/not present). This was a major change for educators at all levels. In addition to ensuring the correct content was present at the right grade/course levels, attention now also needed to be paid to two critical features that impacted how the content was developed: the process standards and the development of conceptual understanding intertwined with skill fluency rather than an exclusive focus on skill fluency.

The Standards called for the newly-named process standards to be incorporated throughout the instructional resources with a two-fold goal. First, the process standards, including problem solving, would be used in the development of mathematical concepts for students; second, the process standards would serve as explicit statements of the competencies students needed in order to "do" mathematics. In addition to the process standards, new attention was to be paid to the development of mathematical concepts, including issues related to coherence across multiple lessons, units, and grades. It also meant attending to the development of deeper understanding by focusing on connections among multiple representations, including an expanded use of mathematical models (NCTM, 1989, pp. 7, 44, 60, 98, 102, 150).

These features were new to most educators, and helping them understand, value, and then adjust their lenses to accurately identify the new features required time and careful attention. Piloting materials with new instructional designs was particularly challenging for both students and teachers who had to adjust expectations and traditional roles. Teachers needed thoughtful support and preparation in order to make fully informed choices. Without this, the innovations in the materials were easily undermined by frustration and misunderstandings on the part of both teachers and students. When managed well, piloting and other activities associated with materials adoption functioned as powerful contexts for professional learning where teachers could begin to internalize the new standards in ways that deepened their understanding and successfully launched the change process across a range of teaching practices.

Discussion: Similarities and Distinctions

It would appear that the challenges that surrounded curriculum resource selection for the innovative curricula of the 1990s are still with us and, in some important ways, have been made even more complex. First, CCSSM replaces the language of the Process Standards with the Standards for Mathematical Practice. However, whether framed as Process or Practice, these standards, which dramatically impact the "how" of teaching and learning, remain perhaps the most challenging features of the CCSSM for teachers to understand, value, and accurately recognize in instructional resources. Many teachers still have little or no vision regarding what the Standards for Mathematical Practice might look like in a classroom or how they might be leveraged to improve learning for students.

Similarly, the focus of the '90s on concept development remains, but is amplified by research coming from studies of the earlier implementation efforts. Today, the CCSSM attempt to push the conversation of concept development further suggests that teachers, not just curriculum developers, will need to understand and actively support the development of a constellation of related concepts over time. These constellations of ideas reach beyond a grade level or course to include the development of mathematical progressions over multiple years. Further, research from the last two decades, such as the body of work from the Quasar projects (Stein & Lane, 1996; Stein, Smith, Henningsen, & Silver, 2009) and the Third International Mathematics and Science Study video studies (Hiebert & Stigler, 2004), has highlighted the role that high cognitive demand tasks, implemented with fidelity, play in promoting student achievement. More recently, cognitive studies are helping us understand that curricular features that promote student engagement and motivation for learning are not all created equal. The effects of "gamified" learning environments present in many of the electronically-based instructional materials are not well understood and are likely to have far reaching implications for learners. These implications for learning need to be understood both as they affect immediate learning outcomes and, in the long term, our responsibilities to inculcate students with the love of learning that can drive personal satisfaction, economic success, and the informed decision making needed for a free and democratic society.

The landscape of learning as it is instantiated in today's instructional resource materials is more complex than it has ever been. Correspondingly, the process of reviewing and selecting new instructional resource materials that incorporate these new features, packaged in technology-based wrappers, is further complicated. The process of instructional resource review and selection must be understood to require careful and well-informed attention. Furthermore, as was the case in the '90s, piloting new resources can be useful, but is enormously challenging for all the old reasons and new ones as teachers and students now also need to prepare for the use of new technologies.

Recommendations Connecting Lessons Learned With Today's Contexts

Given the lessons learned in the '90s and the additional complexity and opportunities in the changing designs of the digitally-based instructional resource materials and revised mathematical standards, three recommendations are offered.

1. *Develop digital resource evaluation tools that make explicit the need to make selections based on a range of important mathematical and pedagogical features.* Selections need to be based in and built upon what we know about powerful and productive teaching and learning, rather than blind enthusiasm for technology-based delivery systems. In addition, once again, it will be critical for educators to have access to resources that will point them toward an examination of new features for which they are likely to have little familiarity. A coalition of well-connected professional organizations (e.g., the National Council of Supervisors of Mathematics, and the National Council of Teachers of Mathematics) would be well-positioned to take on the work of creating such resources.
2. *Create built-in features in new instructional resources that make visible mathematical connections and progressions/learning trajectories for teachers.* Typically, mathematical goals have been presented as a list of discrete skills for each individual lesson. Teachers are left on their own to make sense of the ways in which concepts and skills are designed to come together within a unit of study, and across multiple units, to form a well-connected coherent set of ideas. This work of identifying and making sense of learning progressions that are written into curricula is a particular limitation when educators are working with instructional materials with which they have little or no familiarity (Ferrini-Mundy, Burrill, & Schmidt, 2007). An important and significant improvement would be for authors to create features that would make intentional trajectories and key mathematical connections visible in their materials in support of both the curriculum selection process and the work of implementation.
3. *Make explicit recommendations about preparations needed for effective piloting and develop tools to support productive explorations of new instructional resources.* Given the complexity of reviewing innovative new instructional materials, one way in which developers might help to ensure that reviewers make sense of new features is for authors to create "align-

ment explorations" for educators. These exploration tools would leverage the technology to highlight key features in the new resources, such as the use of the Standards for Mathematical Practice to develop content/concepts, the approaches used to motivate students, content progressions, etc. They might also be used to make explicit the authors' intentions with particular features and highlight new skills or competencies that teachers will need to develop in order to use the materials effectively.

COMPARISON II: IMPLEMENTATION CONSIDERATIONS FOR NEW INSTRUCTIONAL RESOURCES

Lessons Learned

Early adopters of the innovative curricula significantly underestimated the number of years and the initial investment in professional learning time needed for teachers to successfully shift practices in the ways called for in the 1989 NCTM Standards and as intended by the authors of the new resources. The first lesson learned was that changing practice on a large scale in education requires far more than well-designed textbooks. Although aligned instructional resources are necessary, they are not sufficient, and the change needed to implement the new standards required time across many years with accompanying well-designed professional learning experiences (Banilower, Boyd, Pasley, & Weiss, 2006; McREL, 2005).

A second important lesson from the '90s was well documented and described in the Horizon Research studies of the Local Systemic Initiative Projects (Banilower et al., 2006). These studies found that using the new instructional materials as the context for professional learning activities improved teachers' take up and use of new instructional strategies significantly. By using strategically-selected lessons from the new instructional resource materials, specific issues of content, pedagogy, and assessment knowledge could each be addressed with teachers while at the same time familiarizing teachers with the new materials. Earlier, Little (1993) suggested why this approach to professional learning might be effective. He wrote that one key feature of professional learning experiences that can transform beliefs, knowledge, and habits of practice is that these experiences are designed to build teachers' capacity for complex, nuanced judgments about the process of mathematics teaching and learning. Professional learning set in the context of the instructional resource materials provides exactly the type of situation that preserves the complexity of teaching and offers the full classroom context for examination and reflection.

A third relevant finding from the implementation efforts of the '90s was that the degree to which teachers did or did not implement the entire program impacted student achievement. Classrooms where teachers selected a subset of the elements to implement and chose to ignore others saw significantly lower student achievement than those where teachers endeavored to implement the program with fidelity to the intentions of the authors (Briars & Resnick, 2000).

Lastly, in order to help teachers understand and implement the reforms as outlined in the 1989 Standards, professional learning activities needed to *explicitly* name and address changes in both the mathematical content and the pedagogy needed to utilize mathematical tasks built around the components of the standards. The following is offered as an example of such a standard and related task from the middle school level and is intended to illustrate the shifts in the work of teaching that needed to be addressed with teachers. The 1989 Standards called for increased attention to "the connections among a problem situation, its model as a function in symbolic form, and the graph of that function" (NCTM, 1989, p. 126). Tasks such as Tiling Tubs (Figures 7.1 and 7.2) (Friel, Rachlin, & Doyle, 2001) offered teachers opportunities to do just that—connect problem situations to models in symbolic forms. However, tasks such as Tiling Tubs required significant new knowledge and expertise to facilitate effectively. The lesson learned is that it was not enough to simply state these shifts in content—teachers need to see the content in a task, do the task, plan for teaching with the task, and gain experience using the task with students in a variety of settings and over time.

Discussion: Similarities and Distinctions

The lessons learned in the '90s feel as relevant today as they did two decades ago. First, the changes proposed today are no less challenging and will certainly take no less time for educators to understand and incorporate. Kaye Stacey's study of CAS technology implementation in Australia is a dramatic

Tiling Tubs

Hot tubs and in-ground swimming pools are sometimes surrounded by borders of tiles. This drawing shows a square hot tub with sides of length s feet. This tub is surrounded by a border of square tiles. How many 1-foot square tiles, N, are needed for the border of a square pool of edge length S feet?

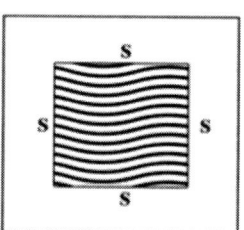

Figure 7.1   Tiling tubs task.

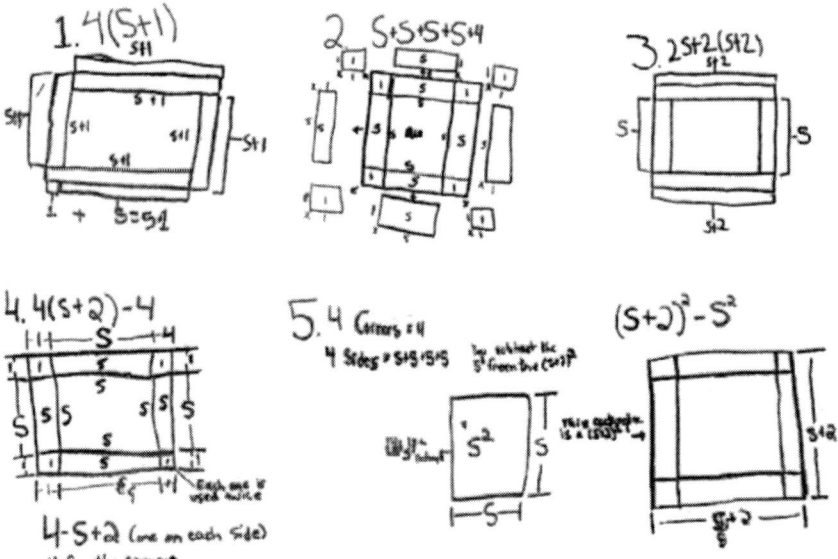

Figure 7.2 Tiling tubs sample student solutions.

and cautionary example of the difficulties that lie ahead as the field attempts to leverage digital resources to support learning (Stacey, 2014).

Second, the additional complexity introduced with the technology-based features in new curriculum resources suggests that professional learning will continue to benefit from being contextualized to lessons within the curriculum. Areas of important study for teachers taking up CCSSM-aligned curricula will still include issues of content, pedagogy, and assessment. However, the framing of the new standards and the evolving understanding of teaching and learning suggest that special attention will need to be paid to learning progressions, the use of high-demand tasks, understanding lesson goals and their connections to prior and subsequent work, and to formative assessment strategies.

It should also be noted that while the changes we will ask educators to make are more complex, the addition of technology also offers possible *curriculum-embedded* professional learning solutions that can ameliorate these challenges. For example, the inclusion of built-in formative assessment tasks similar to the Australian *SMART* Tests design (Stacey, Steinle, Price, & Gvozdenko, n.d.) allows teachers to *both* formatively assess students' thinking and learn about possible categories of misconceptions and solution strategies via the data representations. What interesting possibilities emerge with simply using a carefully-constructed set of assessments that can act as a professional learning opportunity for teachers! A related example is the *Concept Corner* available in the new Pearson online materials (Pearson,

2014) that allows teachers immediate access to Wikipedia-type mathematics content references that have been customized to support teachers using a specific curriculum.

A final example of the need to locate professional learning in the context of the new curricula stems from the fact that digital presentations of curriculum (versus hard copy presentations) allow for more flexible uses of curriculum. Some new curricula, such as *VisualMATH: Functions and Algebra* (Yerushalmy, Shternberg, & Katriel, 2014), go so far as to ask teachers to make decisions regarding lesson and task selection and sequencing. This is a *significant* step forward from the current presentation of instructional materials and would require a deep knowledge of a particular curriculum to thoughtfully consider the affordances of a variety of different options (e.g., do Investigation B before A, or include only Tasks 3–8) for such a flexibly-designed curriculum. It is hard to imagine how a generic discussion of these types of decisions could advantage or even match an exploration of specific choices offered in a particular curriculum. Contextualizing professional learning to the particulars of a new instructional resource to prepare for implementation and to build teachers' content knowledge, pedagogical knowledge, and/or knowledge of assessment continues in this next round of change to be an important professional learning strategy.

In the '90s, we learned that an implementation of a curriculum advantages student learning when it adheres to the intentions of the design team. Widespread access to lessons posted on the Internet and electronically-based curricula increase the potential for supplemental additions and teacher editing by making it easier than ever to change the implemented curriculum from the intended design of the authors (Mills, 2014). Today, the degree of choice in curriculum design ranges from entirely teacher-selected and sequenced at one end of the spectrum, a.k.a. "Curriculum by Pinterest," to limited and scaffolded teacher choice using a defined set of options like the VisualMATH referenced earlier, to, at the other end of the spectrum, a program like Connected Mathematics in which the sequencing and selecting choices are nearly all made for teachers. From "Curriculum by Pinterest" to a research-engineered curriculum design, more study is surely needed to understand the implications for teachers and students across this broad range of options.

Lastly, the understanding gained in the '90s of the importance of professional learning activities that can *explicitly* address the range of knowledge and skills needed by teachers to utilize tasks built around the new reforms continues to resonate as the knowledge base needed expands with the application of technology and higher mathematical expectations. While some of the technology-based tasks and lessons in today's new curricula require similar content and pedagogical content knowledge as their pencil and paper antecedents, such as the Growing Patterns task in Figure 7.3, other

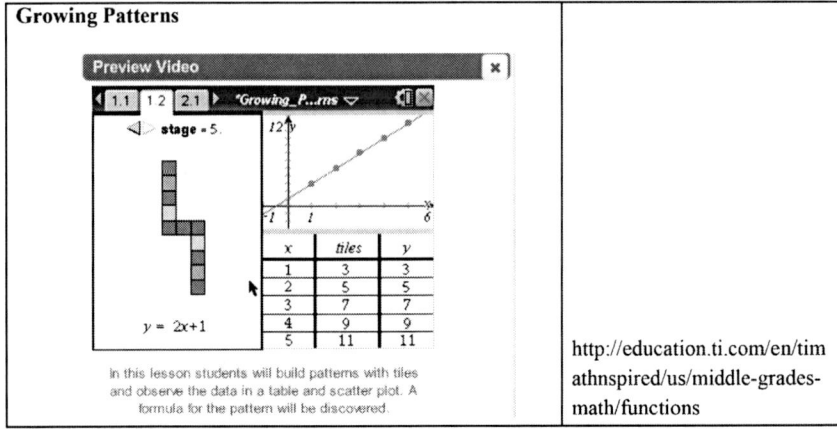

Figure 7.3 Growing patterns.

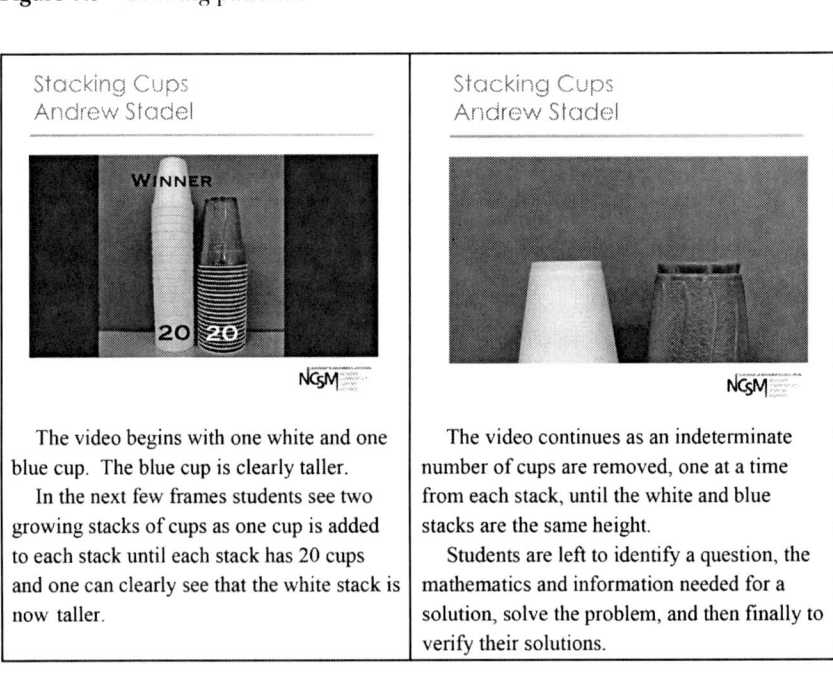

Figure 7.4 Stacking cups.

tasks reach beyond this and will need to be explicitly addressed in professional learning activities (see the Stacking Cups task in Figure 7.4). This seems particularly true as concerns mathematical modeling and the evolving nature of function work from the 1989 Standards to today's CCSSM.

Curricula designed around the 1989 Standards leveraged technology to understand and make connections among representations and to build tables and graphs more efficiently. They also offered students an opportunity to develop a deeper, more flexible understanding of various function classes. The shift today is to require students to use technology to explore change in models as they are portrayed in each representation. Further, exploring real world and mathematical problems to create mathematical models for CCSSM-based curricula also includes asking students to define the problem initially, identifying the variables and mathematical tools needed to solve the problem, and then, in the end, checking whether the mathematical model and solution match the real world application or require a revised model to improve the match.

Recommendations Connecting Lessons Learned With Today's Contexts

1. *Direct more attention toward "Knowledge of Curriculum" in preparation for and during implementation of CCSSM-aligned instructional resource materials.* Perhaps the biggest revelation for me in comparing curriculum innovation in the '90s with today's digital curriculum innovation was the number of ways in which issues associated with Knowledge of Curriculum (Ball & Bass, 2000; Shulman, 1986) surfaced. Consider, for example, the knowledge of curriculum needed to effectively select and sequence lessons and tasks for a course, or the knowledge of curriculum needed to leverage teaching and learning progressions when making instructional decisions, or the knowledge of curriculum needed to effectively incorporate formative assessment into the daily stream of planning and instruction. All of these examples and others shift Knowledge of Curriculum from expertise that is required for curriculum designers and school leaders *toward* expertise needed by teachers if they are going to implement the CCSSM and new instructional resources effectively. With this in mind, curriculum writers and leaders responsible for implementation of new instructional resources will need to pay particular attention to building teachers' knowledge of curriculum.
2. *Give greater attention to explicating mathematical goals for teachers so they understand both the current lesson and the connections to previous and subsequent lessons.* This recommendation builds upon Recommendation 2 (under Comparison 1 above) regarding the need for writers to make visible the intentional mathematical connections and progressions for teachers. Not only is it critical for authors to make this information visible to teachers evaluating these resources for possible adoption, but it is also key to supporting day-to-day planning and instructional decision-making. Only

with this information can teachers make effective decisions regarding the selection and sequencing of lessons and tasks within and across units as well as the numerous instructional decisions required throughout the launch, explore, and summary of individual lessons. Particularly when teachers are just making sense of new instructional materials, working to build a nuanced understanding of the mathematical goals in lessons and units should be at the heart of each and every implementation activity so that it can be at the heart of each and every decision a teacher makes inside the classroom. Making the design of intended goals within lessons and across units visible to teachers is critical for teachers who will be working to use the materials in ways that are consistent with the intentions of the authors.

3. *Find technology solutions to support initial and sustained implementation with fidelity.* As technology innovations come to instructional resources, the recommendation here is to leverage the technology to support implementation. Authors might consider introductory modules for new implementers that could highlight new and/or key features, or draw their attention to critical instructional decisions that are likely necessary to plan for in advance of the implementation. I was struck by the possibilities offered by a "Hive re-sourcing approach" (quasi-crowd sourcing model), as described by Ken Ruthven (2016), that might be a powerful tool for supporting effective implementations by providing a knowledgeable network to collect and distribute information about strategies that are found to be effective and/or ineffective.

CONCLUSIONS

Today's innovation in education built around the widely-adopted CCSSM, a generation of new research findings, and the development of new digitally-enhanced instructional resource materials is stimulating for the field and exciting for the opportunities it offers students and teachers. Viewing today's changes though the lens of lessons learned at a time of similar change beginning 25 years ago affords us the opportunity to avoid the mistakes and build on the successes of that generation. This chapter looked at two aspects of the lessons learned, but other aspects of experiences in the '90s may also provide important perspectives for today's efforts at change.

Large-scale changes of the types currently being offered can be potent opportunities for our community to move closer to the goal of helping every child develop a powerful core of mathematics expertise. Leveraging history's lessons could offer us a proverbial leg up on what will surely be challenging work!

REFERENCES

Ball, D. L., & Bass, H. (2000). Interweaving content and pedagogy in teaching and learning to teach: Knowing and using mathematics. In J. Boaler (Ed.), *Multiple perspectives on the teaching and learning of mathematics* (pp. 83–104). Westport, CT: Ablex.

Banilower, E. R., Boyd, S. E., Pasley, J. D., & Weiss, I. R. (2006). *Lessons from a decade of mathematics and science reform: A capstone report for the local systemic change through teacher enhancement initiative.* Chapel Hill, NC: Horizon Research. Retrieved from http://www.horizon-research.com/the-lsc-capstone-report-lessons-from-a-decade-of-mathematics-and-science-reform/

Briars, D., & Resnick, L. (2000). *Standards, assessments, and what else? The essential elements of standards-based school improvement* (CSE Technical Report 528). Los Angeles: Graduate School of Education & Information Studies, University of California. Retrieved from https://www.cse.ucla.edu/products/reports/TECH528.pdf

Ferrini-Mundy, J., Burrill, G., & Schmidt, W. H. (2007). Building teacher capacity for implementing curricular coherence: Mathematics teacher professional development tasks. *Journal for Mathematics Teacher Education, 10,* 311–324.

Friel, S., Rachlin, S., & Doyle, D. (2001). *Navigating through Algebra in Grades 6–8.* Reston, VA: National Council of Teachers of Mathematics.

Hiebert, J., & Stigler, J. (2004). A world of difference: Classrooms abroad provide lessons in teaching math and science. *Journal of Staff Development, 25*(4), 10–15.

Little, J. W. (1993). Teachers' professional development in a climate of educational reform. *Educational Evaluation and Policy Analysis, 15*(2), 129–151.

McREL. (2005). *McREL Insights—Professional development analysis.* Aurora, CO: Author. Retrieved from http://www.mcrel.org/products-and-services/products/product-listing/01_99/product-68

Mills, V. (2014). Foundations for supporting teachers and the work of teaching. *National Council of Supervisors of Mathematics Newsletter, 45*(1), 2–3.

National Council of Teachers of Mathematics [NCTM]. (1989). *Curriculum and evaluation standards for school mathematics.* Reston, VA: Author.

National Governors Association Center for Best Practices & Council of Chief State School Officers. (2010). *Common Core State Standards for Mathematics.* Washington, DC: Authors.

Pearson. (2014). *Pearson System of Courses: Getting Started Guide* (PowerPoint slides). Retrieved from http://support.pearsonschool.com/default/assets/File/Getting%20Started%20Guide%202014-07-25.pdf

Ruthven, K. (2016). The re-sourcing movement in mathematics teaching: Some European initiatives. In M. Bates & Z. Usiskin (Eds.), *Digital curricula in school mathematics* (pp. 75–86). Charlotte, NC: Information Age.

Shulman, L. S. (1986). Those who understand: Knowledge growth in teaching. *Educational Researcher, 15*(2), 4–14.

Stacey, K. (2014). *Mathematics curriculum, assessment and teaching for living in the digital world: Computational tools in high stakes assessment* (PowerPoint slides). Retrieved from https://s3.amazonaws.com/csmc/CSMC_STACEY_CAS_edited.pdf

Stacey, K., Steinle, V., Price, B., Gvozdenko, E. (n.d.). *SMART: Specific Mathematics Assessments that Reveal Thinking* (General Information). Retrieved from http://www.smartvic.com/smart/index.htm

Stein, M. K., & Lane, S. (1996). Instructional tasks and the development of student capacity to think and reason: Analysis of the relationship between teaching and learning in a reform mathematics project. *Educational Research and Evaluation, 2*(1), 50–80.

Stein, M. K., Smith, M. S., Henningsen, M. A., & Silver, E. A. (2009). *Implementing standards-based mathematics instruction: A casebook for professional development* (2nd ed.). New York, NY: Teachers College Press.

Yerushalmy, M., Shternberg, B., & Katriel, H. (2014). *VisualMATH: Functions and Algebra* (Instructional materials). Retrieved from http://visualmath.haifa.ac.il

CHAPTER 8

TECHNOLOGY TO SUPPORT MATHEMATICS INSTRUCTION

Examples From the Real World

Loretta J. Asay

Clark County School District provides services to over 320,000 students in over 8,000 square miles around Las Vegas, Nevada. The district has 357 schools and more than 22,000 teachers. Within the school district, the Instructional Technology program is coordinated as part of the Instructional Design and Professional Learning Division, providing services to all schools, teachers, and students. Information about Clark County School District can be found at http://ccsd.net.

The project highlighted in this chapter is known as the e3 Project and provides over 15,000 devices to students, for one-to-one technology (in which every student receives a device), 24 hours a day and seven days a week. Schools choose the devices, either iPads or small laptops known as netbooks. As e3 is funded through Title I allocations, the participating schools all have at least 75% of their students in poverty. As well, they all have large proportions of students whose first language is not English.

"e3" refers to "engage, empower, and explore" and was designed using research done by the Project RED (Revolutionizing Education) group. In *The Technology Factor: Nine Keys to Student Achievement and Cost-Effectiveness* (Greaves, Hayes, Wilson, Gielniak, & Peterson, 2010, p. 12), the following nine factors are identified as predictors of student learning with technology:

1. *Intervention classes*: Technology is integrated into every intervention class period.
2. *Change management leadership by principal*: Leaders provide time for teacher professional learning and collaboration at least monthly.
3. *Online collaboration*: Students use technology daily for online collaboration (games/simulations and social media).
4. *Core subjects*: Technology is integrated into core curriculum weekly or more frequently.
5. *Online formative assessments*: Assessments are done at least weekly.
6. *Student-computer ratio*: Lower ratios improve outcomes.
7. *Virtual field trips*: With more frequent use, virtual trips are more powerful. The best schools do these at least monthly.
8. *Search engines*: Students use daily.
9. *Principal training*: Principals are trained in teacher buy-in, best practices, and technology-transformed learning.

The e3 Project exists to transform teaching and learning in classrooms through access to current technologies. When we refer to "transforming," our district uses two models. The first, known as SAMR (see Figure 8.1), is the work of Dr. Reuben Puentedura (2013). This model is a way of classifying how technology is used for learning. In the *Substitution* level, there is no functional change in the task; it is simply being done through technology. Students receiving information from a digitized body of text would be an example of substitution. The next level is *Augmentation*. In this type of activity, the technology is a substitute for previous types of tasks, and there is an improvement in the task because technology is being used. When a digitized body of text, for example, includes a way for students to highlight passages, take electronic notes, or mouse over for a pop-up definition, this is a functional improvement in approaching the task. In the third level, *Modification*, the fundamental task is changed. Graphing calculators, for example, modify the task, as students can quickly create graphical representations and change them easily. This direct manipulation of the data is a different activity than either looking at examples or graphing by hand. Finally, at the *Redefinition* level, the task has been redesigned as a new task, not possible without the technology. As an example, tasks in which students determine what data to gather, then gather it through probeware or electronically collaborative methods, then represent it and share it, are impossible without

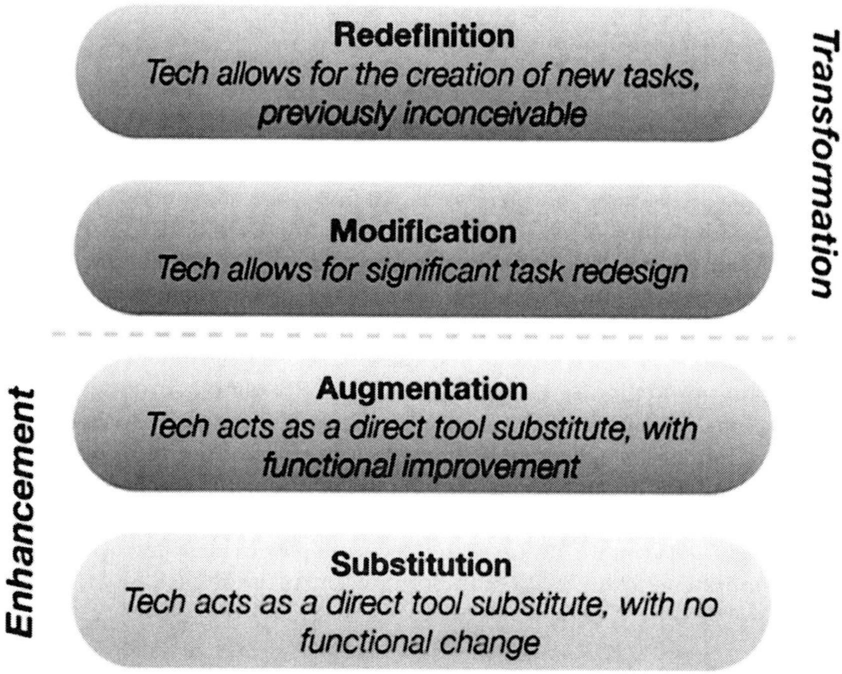

Redefinition
Tech allows for the creation of new tasks, previously inconceivable

Modification
Tech allows for significant task redesign

Augmentation
Tech acts as a direct tool substitute, with functional improvement

Substitution
Tech acts as a direct tool substitute, with no functional change

Transformation

Enhancement

Figure 8.1 Four levels of SAMR model.

the technology. Moreover, such tasks are fundamentally different than nontechnological attempts at the same types of tasks (Puentedura, 2013).

This classification of activities is meant to serve as a framework. As a model, the labels can be argued about for any task; the lines blur between modification and redefinition, for example. I have found that the value comes in the conversations about those tasks, not in an absolute agreement on the labels.

The second framework important in the e3 Project is the Technology Integration Matrix or TIM. This matrix provides a way of looking at aspects of a rich classroom (active, collaborative, constructive, authentic, goal-directed) and where the classroom is for each aspect along a continuum from adoption to transformation. It is not expected that any classroom will be constantly in the transformation range on all five aspects; it is expected that we will see a shift to the right of the matrix for the five aspects over the course of time as technology is integrated (Florida Center for Instructional Technology, n.d., *Technology Integration Matrix*). There is an accompanying observation protocol, TIM-O (Florida Center for Instructional Technology, n.d., *Technology Integration Matrix Observation Tool*), which we use for making classroom observations and guiding conversations with administrators and teachers during professional development.

WHAT WE FOUND

Over the course of the last three years, we found four factors that influence the transformation of classrooms: incorporation of blended learning (a mix of face-to-face and digital instruction), access to information, student demonstration of understanding, and robust professional development. These factors emerged from analysis of survey responses from e3 teachers at the end of each school year. Responses were grouped and coded according to themes. In the spring of 2014, for example, over 80% of the e3 teachers responded to the survey questions. We isolated responses from the mathematics teachers, and their identification of the most important student benefits became visible. Among the mathematics teachers, access to information and tools, increased collaboration (a component of blended learning), and demonstration of understanding were identified as the most important benefits by 71.7% of the respondents (see Table 8.1).

Blended Learning

Because each student has a device, teachers can use flipped and rotational models of blended learning (Clayton Christensen Institute, n.d.). When teachers flip instruction, they use digital content, such as teacher-made or teacher-gathered videos, rather than traditional direct instruction. When students do this outside of class time, the teachers find they can use that precious

TABLE 8.1 Most Important Benefits of One-to-One Devices According to Mathematics Teachers

Benefits	Percentage of Responses
Access to information, such as Internet resources, teacher-created resources, notes, portfolios	32.6
Student attitude improvement, including increased engagement and taking responsibility for learning	19.6
Increased interactions and collaboration, teachers with students and students with students	15.2
Access to mathematics tools	15.2
Students demonstrate understanding of concepts	8.7
Other, including student choices, practice, and projects	8.7

instructional time in class differently, for projects, for individual help, for group work, and for tackling difficult assignments students struggle with when the teacher is not around. About 30% of the mathematics teachers used a form of flipped instruction at least once during the 2013–2014 school year.

When direct instruction happens electronically during class, we consider this a rotational or lab model. Usually, the class is using a purchased digital curriculum, allowing students to move with individual pacing. In these classrooms, teachers often set up rotations among group work, digital work, and teacher-assisted work. About 40% of the mathematics teachers used a rotational or lab model for at least one unit during the 2013–2014 school year.

Internet Access

When we first introduced teachers to using the Internet with instruction, the biggest concern was that students would not have Internet access outside of school. We formed partnerships with local Internet providers for low-cost access if families chose to subscribe. Over 85% of our students reported having access when surveyed in the spring of 2014.

Changes

Teachers report that both blended learning models, and all their variations, require a lot of work from the teacher. They cannot sit back and expect a digital curriculum or instructional videos to do all of the teaching. Our teachers who have embraced these methods see their roles as both redefined and critical.

Students report that they like working this way. They find it engaging. Through software, teachers are able to see how long students work at tasks, how often they repeat a video, and even what types of questions they pose to their peers. At the same time, it is a culture shift for the students. One young man told me, "It sucks. I have to work more."

Access to Information and Tools

Internet

Access to information took on many forms. First, students had constant access to Internet resources and the ability to conduct research to find answers and explanations. In a parent focus group, one mother told me, in broken English, "For you it may be a little thing. But now my daughter has the world in her hands." This was her way of expressing, as did most of the parents, that access to information was important for their children.

Teacher-to-Student and Student-to-Student

Second, students and teachers had electronic access to each other. One of the most popular tools was Edmodo, an online learning management system into which teachers and students can put resources and information, assignments, blogs, and quizzes. Students can collaborate with each other, with the work visible to their teacher. The teacher can work individually with a student, with groups of students, or with all students. For access to the protected work space, each student must have a code given to them by the teacher. One of our teachers made a rule that she would no longer answer student questions posted in Edmodo after 9:30 p.m. She then found that most of her students simply started helping each other. Two of our mathematics teachers set up a homework space in Edmodo, but did not provide homework help. Instead, students asked for the code to the space and then helped each other. This homework space is one of the most active electronic groups in this project. With a device for each student, it is simply impossible, and unproductive, to halt student collaboration.

Resources

Finally, students had constant access to resources. Some, such as videos or notes, were posted by their teachers. Some were organized and filed digitally by the students themselves. The construct of "notes" took on new meaning. Notes now included video, links, photos, homework, and examples. In addition, students quickly found they could search for information they had stored. It was very common to walk into a classroom and see students snapping photos of work on the board, using note-taking software, and electronically exchanging notes or information. Successful teachers were strategic about classroom management; less successful teachers complained frequently about the flow of student communication. Strategies, such as signals for turning the devices face down and signals for listening to the teacher, were necessary.

Background Knowledge

Seldom mentioned by teachers, but now a focus of our professional development, was access to information that could build background knowledge. We know that students learn by accessing their background knowledge; however, their background knowledge is not always what we expect or assume it to be (Fisher, Frey, & Lapp, 2012). For example, I watched a lesson in which percentages were being used in the context of mortgage interest. As I looked around the room, I saw confusion on many students' faces. The teacher lectured on, oblivious to the fact that few of her students had any background knowledge about mortgages. I realized that most of these students lived in apartments; neither owning a house nor a mortgage were part of their experience. Banks and lending institutions were also not familiar to them, for that matter. Especially given the diversity of our students in the United States, we must tap into digital resources for helping students build background knowledge that will help them learn.

Demonstration of Understanding

Teachers discovered that one-to-one devices allowed students to demonstrate their understanding frequently and easily. Electronic tools were used for both formative and summative assessment.

Closure
Closure is an important part of the lesson cycle and can provide an opportunity for formative assessment (Grouws, Tarr, Sears, & Ross, 2010). Teachers developed a variety of strategies for "exit tickets," closure activities in which students leave artifacts that demonstrate understanding of the lesson. Often, these were in the form of video or photos. For example, using screencasting software, a teacher would give a mathematics problem and ask the students to record themselves solving it, narrating their process. These were then posted so that the teacher could review them later. It was eye opening to have access to student understanding, misconceptions, and mistakes. Teachers were then able to adjust instruction, group students, or re-teach based on these exit tickets.

Summative
Teachers also found a variety of tools to use for electronic quizzes and tests. The best of these were also seen as valuable for preparing students for the Smarter Balanced Assessment Consortium (SBAC) tests of the Common Core State Standards for Mathematics (CCSSM; National Governors Association Center for Best Practices & Council of Chief State School Officers, 2010). Teachers preferred those that allowed teacher flexibility in setting up and/or choosing questions, rather than drawing on pre-made banks of quizzes and tests that often came with their textbooks.

Professional Development

Job-Embedded Coaching
In the one-to-one project, we provide instructional coaches (Knight, 2007) who facilitate quarterly workshops for the teachers. At each workshop, each teacher set a goal for how the devices will be used for the next quarter, based on the theme of the workshop. For example, in the quarter one workshop, the focus was on collaboration, and participants worked together to develop ways students could collaborate more, using the iPads. Each participant chose at least one activity, strategy, or project that would help students collaborate more and committed to implementing it. The instructional coaches then followed up with each teacher, helping him or her to choose resources, practice, co-teach, and document the experience.

On-Demand Professional Development

In addition to the coaching and workshops, we found that teachers needed on-demand assistance as they planned for instruction. This was the basis for Bringing Learning and Standards Together (BLAST), a state-funded collection of online modules for each of the CCSSM.

Each module is designed to take 20 or 30 minutes for a teacher to complete. Each consists of background information about the standard or cluster of standards; assessment information; ideas for instruction, including how to notice and address misconceptions and suggested discussion questions; and a space for collaborating with colleagues. Teachers often asked, "What does it look like in a classroom?" when the CCSSM were introduced in our state. We made sure that modules included classroom video, student work samples, or interviews with teachers.

Mathematics teachers in Nevada access the modules through an online curriculum planning tool known as the Curriculum Engine. This is a repository of all vetted curricular resources for Nevada teachers, organized according to content standards. The Curriculum Engine contains detailed background information about each standard, resources to support classroom instruction, examples of student work, and assessment strategies. As teachers are planning for instruction, they have the option to dig deeper into the standards through the BLAST modules. Several schools have also used various modules as the basis for face-to-face professional development (the BLAST modules are available at http://blast.ccsd.net).

SUMMARY

One-to-one projects are seen less and less as projects and more as an expectation in schools. Students often carry devices capable of letting them collaborate, create, and research. In the world of K–12 mathematics education, we need quality and easily accessible resources that tap into these 21st century skills our students use.

REFERENCES

Clayton Christensen Institute (n.d.). *Blended learning model definitions.* Retrieved from http://www.christenseninstitute.org/blended-learning-definitions-and-models/

Fisher, D., Frey, N., & Lapp, D. (2012). Building and activating students' background knowledge: It's what they already know that counts. *Middle School Journal, 43*(3), 22–31.

Florida Center for Instructional Technology (n.d.). *Technology integration matrix.* Retrieved from http://fcit.usf.edu/matrix/download/tim_table_of_summary_indicators.pdf

Florida Center for Instructional Technology (n.d.). *Technology integration matrix observation tool.* Retrieved from http://fcit.usf.edu/matrix/tim-o

Greaves, T., Hayes, J., Wilson, L., Gielniak, M, & Peterson, R. (2010). *The technology factor: Nine keys to student achievement and cost-effectiveness.* MDR.

Grouws, D. A., Tarr, J. E., Sears, R., & Ross, D. J. (2010, January). *Mathematics teachers' use of instructional time and relationships to textbook content organization and class period format.* Paper presented at the Hawaii International Conference on Education, Honolulu, HI.

Knight, J. (2007). *Instructional coaching: A partnership approach to improving instruction.* Thousand Oaks, CA: Corwin Press.

National Governors Association Center for Best Practices & Council of Chief State School Officers. (2010). *Common Core State Standards for Mathematics.* Washington, DC: Authors.

Puentedura, R. (2013, May 29). *SAMR: Moving from enhancement to transformation* [Web log post]. Retrieved from http://www.hippasus.com/rrpweblog/archives/000095.html

CHAPTER 9

WE THOUGHT WE KNEW IT ALL

Josephus Johnson

This is a story of how my school adopted technology. It is a story about tempered optimism and undaunted hope. It is a story about setbacks and big achievements. It is a story about how we fumbled and bumbled our way through the opening of a brand new school.

Battle High School is a school of about 1400 students in Columbia, Missouri. Although Columbia is a typical college town with many highly educated people and very successful businesses, Battle draws from the north side of town where much of that affluence has failed to penetrate. Our population is heavily burdened with poverty and all the issues that accompany it. These issues were always at the forefront of our planning as we prepared to open our brand new building. From start to finish, it took six years to plan Battle High School. Although focusing on the success of our students was always a priority, there were other issues that were unique to our school.

One of the issues that poverty brings with it is the lack of access to information. In an effort to combat that issue, our district decided that Battle High School was going to be the first school in our district to go 1:1 with technology. We spent several months of committee meetings trying to decide which type of device we wanted to employ. After considering many factors, including costs, usability, battery life, durability, and others, iPad minis were chosen. Although this empowered our school to move education to a new frontier, it also gave us another facet of our building to which no one else in our town was capable of relating.

A third facet of our new school that complicated our journey was the faculty. When our school was opened, that year not only marked a new building in our city but also a realignment of our whole school system. Before 2013, we had a K–12 system with 4 buildings: Elementary School was K–5, Middle School was 6–7, Junior High School was 8–9, and High School was 10–12. The administration realized that our students were undergoing too many transitions, especially during the teen and pre-teen years where at-risk students are likely to disengage from school. The new alignment for Columbia was to have just 3 schools where students transition to Middle School in 6th grade and High School in 9th grade. To staff these buildings, all the building principals met for what was dubbed "teacher draft day." At the end of that meeting, some buildings had a handful of new staff members while others had over 70% new staff members. Each building had their own set of difficulties for community building with staff. But Battle once again was more difficult since we had a completely new set of teachers, secretaries, guidance staff, janitors, etc. Each of us came with a very clear understanding of what we had done in the past at our old schools, yet lacking a clear vision of what Battle High School was going to be and the trust needed to enact any vision.

These three challenges brought together a set of problems that crystalized into this question: How could we unite a group of teachers who didn't really know each other to develop technological methods that were still just hypothetical to motivate, teach, and inspire a group of students who had a multitude of reasons not to be motivated, taught, or inspired? Now fast-forward through our first year and all of the wondrous experiences, and I am here to tell a bit about what we learned.

To begin, I will be optimistic and talk about all the things that we were right about. Teacher development was a huge part of our plan and we took the "all of the above" approach. We had sessions about every aspect of technology that we could contemplate and differentiated the sessions for different teacher levels of experience and comfort with technology. We attempted to address every issue and opportunity that we could foresee, and it did allow us to start off on the right foot. One of the parts of the teacher professional development was the SAMR (Substitution, Augmentation, Modification, Redefinition) model (see Asay, 2016) that identifies four different layers of technology integration. We encouraged all our teachers to at least become comfortable with the first level, *substitution*, where student work would be relatively unchanged, but the work would be digital instead of on paper. This would include things like making class notes available online, having digital worksheets, and collecting assignments electronically. These were all easy changes and would not result in large adjustments in the way a class "felt." We were hopeful that future changes would be more substantive, but this tactic would ensure all teachers would have the tools to use their iPads on any day.

The last piece of what we got right was choosing the right type of person to facilitate teacher professional development. We wanted someone who could effectively lead a large group in a conference room, comfortably interact one-on-one with a teacher in a classroom, and informally interact with a group of teachers in an unstructured place like the teacher lounge. That list of requirements was hard to fill, but finding the best people to get in front of our teachers to lead the way through the minefield of technology was important to a smooth transition.

The leadership team of our school was quite happy that almost every area we identified as a concern was dealt with and the staff handled it appropriately. Certainly, there are areas that we wish to continue to build upon. Every week, we identify and celebrate teachers who are discovering some new strategy to help students use iPads in a transformational way. We find ways to coach and empower our students to move beyond the limitations of their community. We continue to plan activities that build the staff community to craft our vision of what it means to be a teacher at Battle High School. Like any other school out there, it is not just a matter of building success; it is a matter of maintaining the commitment to success for each student and teacher in the building.

We made several good decisions with the plan of how we would open Battle High School, but there were also several things that we were not able to anticipate. The first issue we faced was the lack of apps for learning high quality mathematics tasks in high school. This has been improving slightly in the past months, but we are still looking for some apps that promote high-level cognitive demand tasks in a way that is "cool" for kids. We have found a couple that work for limited objectives. 123D Make is amazing for analyzing rotational solids in Geometry and Calculus but useless elsewhere. There are several digital Learning Management Systems (such as Blackboard, Showbie, Schoology, and others) that are great for digitally passing assignments between students and teachers, administering digital quizzes, and other logistical issues within a classroom. However, they are designed as classroom aids, not as tools to help teach or learn rich mathematics. Desmos is an amazing graphing calculator but does little else. Geogebra is great for exploration into ideas but has a steep learning curve for many students and teachers. While many platforms are improving, the Holy Grail of digital learning guides has yet to be created.

Another problematic area has been that some students can be reluctant to delve into some digital environments. Our principal often talked about the students as "digital natives" and how they would be the easiest converts to a 1–1 school, but she failed to address the aspect of a learning curve for productivity software. While there were practically zero students afraid of the devices (as opposed to some of the terrified teachers), they were not always quick to embrace email, productivity apps, and some of the professional aspects of the iPad.

A third area where our crystal ball disappointed us was how difficult it would be to get teachers to move beyond the iPad as paper replacement. Our group did a great job of adopting technology in the beginning, but others have stalled to get further into the possibilities of what iPads can do. Allowing kids to make videos, take voice notes, or collaborate digitally would be relatively easy adjustments for teachers to investigate, but there are always reasons to not move forward. Some teachers were frustrated by technology failures. Our wifi would occasionally go down and, although this only happened maybe 8–10 times all year, inevitably it happened when a technophobe was trying something new. Most of our teachers were resilient enough to persevere through all the speed bumps, but we also found some teachers didn't have the professional fortitude to persevere through the small issues.

One more subtle area of note is something that I have been forced to encounter as a function of my position as department chair. I think it is best illustrated by a recent encounter with my head principal. We had a meeting about the progress of the new teachers in our building. She praised many of the teachers for their commitment to help each student in their classes. That particular issue is a significant one for her as she never gives up on any of her students. No matter how far along the year has gone, she always feels there is room to make a difference for her kids. And she truly means EACH kid. In her mind, she would give just as much help to an Algebra I student who hasn't turned in an assignment all year as she would to an AP Calculus student who has spent an hour after school each day. That dedication struck me when I heard her formulating a judgment about staff members before they even had a full semester of teaching in our school. I was struck by the deep divide between how far she would work for her students and how little she would fight for her staff. It reminded me that, as a building leader, I have to work with all the teachers and all their faults. I have to help the teachers who want to use technology every day and believe that all kids will grow to do great things. I have to work with the teachers who are overstretched with committee work each day. I have to work with teachers who are coaches and have no time to work on lessons for 5 months each year. I have to work with teachers who have families of their own and can't take time to stay after school to help students. I also have to work with teachers who don't believe their kids can succeed. Just like my principal feels a deep connection to and passion for each of the 1400 students who attend her school, I have a commitment to make each math teacher promote a better environment for student learning. This applies to the new teachers who don't know, the veteran teachers who do know, and everyone in between. As a department chair whose power and influence only extends to the walls of my building, I have to find a way to empower and educate all the teachers because that is the only way to help all of the kids.

A final issue that was not fully anticipated resulted from our decision to allow all of our students to have full use of their devices without limitations. Our optimism told us that young adults would live up to our expectations and keep education as the primary focus during school hours. But we were sorely mistaken. A large majority of our kids took the freedom of an unlocked iPad and made bad choices. Games, Internet surfing, and social media have become serious distractions for our population beyond the normal teenage issues. We tried to educate and inform our students about the dangers of digital distractions, but the temptation was too strong for most students. Even for adults, it is difficult to ignore a cell phone ringing in your pocket. Asking students to focus on math and finish their explanation before checking the picture their friend just sent them has proven to be a Sisyphean task. I have expanded upon this issue because I think it is exacerbated by the nature of the iPad. It was conceived and designed by Apple as a media consumption device and, although it does great things in other realms, that is the sweet spot for usage. It takes a lot of student willpower and a fair amount of teacher technical expertise in creating experiences to make it so that students can work uninterrupted on their devices during the school day, and most students are lacking in those areas.

Finally, I ask for a few moments to express my wish list of things that I want for my school. I think that we need more content tools available to us. Textbooks and ancillary supplies of olden days were easily adjustable and personalized by teachers. It would be nice if we could find a way for the future of digital materials to give a strong base that allows personalization by the teachers, but also keeps the cohesion of a professionally developed curriculum. We also need opportunities to try out many different varieties of teaching and learning in the digital environment. Few people have a clear idea of what a "good" digital classroom looks like. We have lots of jargon about flipped classrooms, customizable learning, facilitated student collaboration, and so forth. However, when you push people to describe what that means in detail, our profession has not been able to give a clear answer. Teachers are going to need time to attempt and fail. We will need the support of start-up companies to spread ideas. We will need university support to push individuals to perfect their programs for all students in all areas of the world. That pattern will be fraught with failures and false starts, but our industry needs to figure out what "good" digital learning is really all about.

REFERENCE

Asay, L. (2016). Technology to support mathematics instruction: Examples from the real world. In M. Bates & Z. Usiskin (Eds.), *Digital curricula in school mathematics* (pp. 123–131). Charlotte, NC: Information Age.

CHAPTER 10

DEEPLY DIGITAL CURRICULUM FOR DEEPLY DIGITAL STUDENTS

Brian Lemmen

*Tell a child WHAT to think, and you make him a slave to your knowledge.
Teach a child HOW to think, and you make all knowledge his slave.*
—Henry A. Taitt, 1982

The above quote by Henry A. Taitt is one of my favorites. Are students' minds empty vessels that we as teachers need to fill? Or, are students naturally curious and interested in finding answers to problems? I am in my 32nd year of teaching middle school and high school mathematics. I started my teaching career by filling students' minds with mathematical procedures where I did most, if not all, of the thinking for them. I taught mathematics as a list of topics and procedures with no context in which to apply them. Looking back, it was as if I attempted to make students puppets. Teaching this way stifled exploration and original thought, and true learning did not take place. Today, I try to get my students to see mathematics as a tool that can be used to help them better understand the world around them. My philosophy of teaching mathematics has changed considerably over the years. There are two phenomena that account for this. They are the birth of reform mathematics curricula and the emphasis on using technology to help teach mathematics. Both of these allow us to use mathematics as a tool to explore the world around us and to let student learning occur. Before I get into the use of the "Deeply Digital" curriculum, some background information is needed.

In 1996, after thorough reviews of many mathematics curricula, Holland Christian High School in Holland, Michigan, adopted the first edition of the Core-Plus Mathematics (CPMP) curriculum. Once it was adopted, I remember my feelings of anxiousness, nervousness, and uneasiness. For the first time in my life, I had to *think* mathematics. I could "do" mathematics, but I was never really challenged to think or apply mathematics. With this adoption, my pedagogy for teaching mathematics changed. For the first time in my teaching, I witnessed students thinking mathematically and applying the mathematics they had developed.

In 2006, Holland Christian High School implemented a one-to-one laptop initiative where every student makes use of a laptop throughout his or her school experience. We have enjoyed the challenge of using technology to help us teach more effectively. This technology-rich learning environment has gained the attention of experts around the United States as we received the "Apple Distinguished Program" award in 2013–14 and 2014–15 for our demonstrated success in providing an exemplary learning environment and being a center of innovation, leadership, and educational excellence. The administration strongly encourages the teachers and staff to be innovative in the use of technology in our disciplines. However, the mathematics department lagged behind other departments in the use of technology. We were using technology really to support what we were doing rather than using technology to revitalize and change how we taught.

The adoption of the Core-Plus Mathematics curriculum led us to become a pilot school for the second edition. Soon after that, the Michigan State Department of Education made four years of mathematics a requirement for graduation. This led the teachers to think carefully about an alternate fourth-year course for students who were going to college but perhaps not into a field requiring calculus. With this in mind, we took an opportunity to become a pilot school for Transition to College Mathematics and Statistics (TCMS) that provided our school with a needed course option for seniors. TCMS is a problem-based, inquiry-oriented, fourth-year high school mathematics course intended for students who do not plan to major in mathematical, physical, or biological sciences or engineering. It is a National Science Foundation-funded curriculum.

Because of the high school's involvement as a pilot site with secondary mathematics curriculum projects and the one-to-one status of our school, we were provided a unique opportunity to work with AJ Edson, then a graduate student at Western Michigan University. AJ redesigned an instructional unit from the TCMS curriculum that we were using, called Binomial

Distributions and Statistical Inference, and turned it into a "Deeply Digital" curriculum (see Edson, 2016). The use of the digital delivery of the curriculum materials was like upgrading from a minivan to a Lamborghini.

Trying to describe a Lamborghini is difficult. You could use terms such as awesome, fast, exquisite, explosive, and so forth. Trying to describe "Deeply Digital" is similarly difficult. Any words or phrases that you would use to describe the experience of driving a Lamborghini could similarly be used to describe the use of the "Deeply Digital" curriculum. My research into Lamborghinis revealed a six-speed transmission. I would like to describe six features of the digital version of TCMS. I describe how each feature made a difference for my students and how each feature changed the way I needed to think about my teaching. As we move through the gears, we will examine the affordances that the Deeply Digital version offers for both the students and the classroom teacher.

GEAR 1: A BRIEF OVERVIEW OF THE STUDENT INTERFACE

The digital shell is elegantly laid out. There is a unit home screen that provides an interactive table of contents for the multi-day investigations that make up each of the four lessons of the unit (see Figure 10.1). The

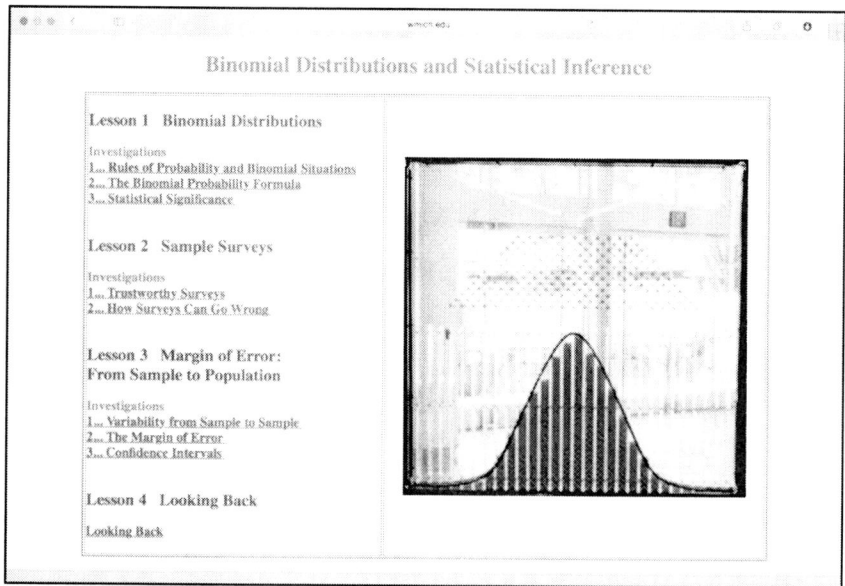

Figure 10.1 Home screen and unit opener of the digital materials.

interface makes navigation easy for students and the teacher. About two clicks gets you access to most places and features of the digital materials.

Once a user selects an investigation, the digital materials are divided into four main window frames (see Figure 10.2). The largest, along the left-hand side, is what we would think of as the print material in a conventional book. A second window frame to the upper-right is a place for students to do their work and communicate with each other. The third window frame to the lower right is where embedded videos and audio tracks can be played and where TCMS-Tools, a suite of mathematical and statistical software developed for the curriculum materials, can be used. Along the top is a fourth frame that is a navigational toolbar with a home button, navigation arrows, TCMS-Tools, text-to-speech software, a glossary of key terms, and a button that creates a place for students to write notes.

It was fascinating and exciting to watch my students navigate easily using the Deeply Digital materials, with little instruction the very first time. Because of the elegant layout and ease-in-use functionality, there were hardly any questions that emerged from students about navigating the digital materials.

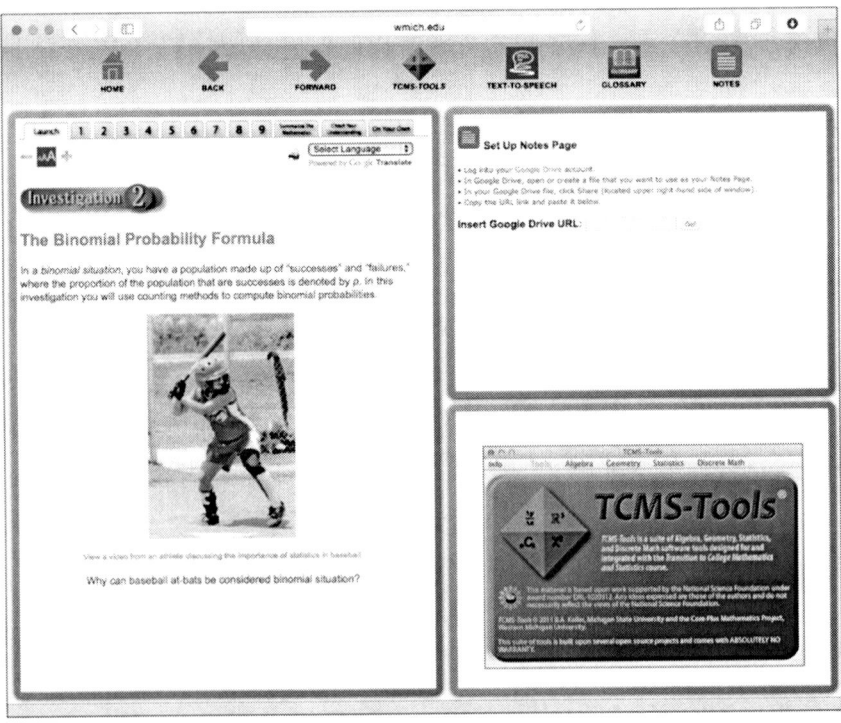

Figure 10.2 Main components of the digital materials.

GEAR 2: THE "UPSIDE DOWN" QUESTIONS AND STUDENT USE OF DIGITAL SCAFFOLDING

AJ did not simply take the unit, make it a PDF version, and call it Deeply Digital. It was much more than that. He redesigned the way problems were delivered. The TCMS print material has scaffolding that helps guide students to desired conclusions. Students using the print material work in groups and discuss the questions together leading up to the final question, the heart of the matter, the big picture idea. What AJ did in the digital materials was to ask the big picture question first. He did not provide the scaffolding at the beginning. He turned the questions "upside down." At first this may not sit well with you. But if you philosophically believe that students are curious and have the ability to reason, it is a prodigious approach. Too often teachers underestimate students' abilities.

An example of how problems were transformed from the print format to the digital version can be seen in Figure 10.3. This problem is taken from a lesson in which the addition and multiplication rules of probability are being reviewed and extended. In this example, one of the big ideas being explored is investigating the different ways students can find probabilities, including finding the probability of at least one success. For example, P (at least one win) = $1-P$ (lose all 10 times). In the digital materials, AJ asked the big picture problem first without the scaffolding. But if you look closely, there is support provided for students in the digital format. If a student needed some help or desired some scaffolding to think about the big question, there were "blue buttons" that students could click as they felt necessary. When clicked, the links provided assistance in the form of another problem. So the scaffolding-type questions were available, but only if the student needed them. This made the curriculum interactive and gave the student choice. Giving the students choice was a very important difference between the two formats. The print version made assumptions that students would need help. It reminds me of my early philosophy of teaching. Recall that when I started teaching, I did most of the thinking for my students. The built-in scaffolding in the print version directs students' thinking. The digital materials better align with how I think about good teaching practices today. I want students engaged, interacting, and thinking with the help of their group members. Because students are asked the big picture problem first in the digital materials, without the preliminary scaffolding the print text offers, the students were immediately engaged and required to think more deeply. All along, students know they have the choice of scaffolding that the blue buttons offer.

Use of the digital materials changed the way my students engaged with the mathematics and also changed the way I engaged with my groups. When I engaged with groups in the print version, I had to ask what scaffolding

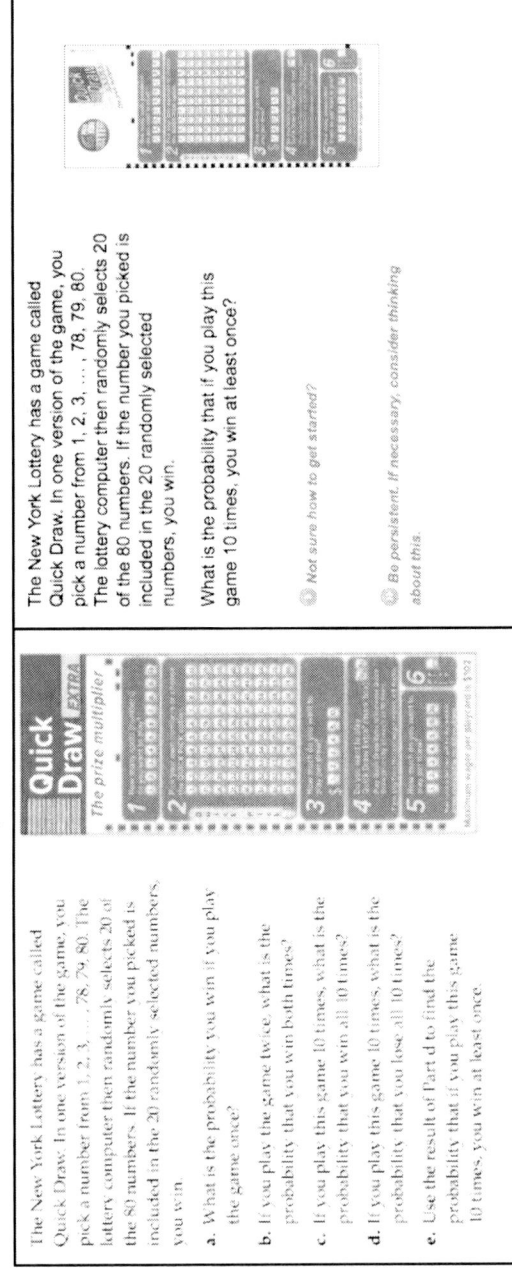

Figure 10.3 Print and digital delivery of an example problem.

question they were struggling with and why. In some instances, because they were falling a bit behind other groups, I would have them skip scaffolding questions and move on to the big idea question. Because the big idea question in the digital version was presented first, I had to ask, "Did you push the blue buttons?" If the answer was no, I suggested they do so and returned after students considered thinking and discussing the prompted problem. The blue button questions provided needed scaffolding for struggling students. This, however, was not their only purpose. I learned that, even if the students did not feel it necessary to push the blue button, it was a good idea to suggest they do so. Pushing the blue button after they had completed the problem either reinforced their thinking or caused them to reengage with the problem. It was truly fascinating watching the development of this dual purpose for the blue button questions.

GEAR 3: THE LAUNCH OF A LESSON OR INVESTIGATION AND USE OF VIDEO AND AUDIO CLIPS

The *Launch* is the first of four phases of the TCMS instructional model. In both the print and digital versions, a launch happens at the beginning of each lesson. The launch is followed by a second phase, *Exploration*, which involves students working in groups answering questions related to the launch situation. Phase three is a time for *Summarizing* and *Sharing* key ideas learned or explored in the previous phase. Finally, there is a time for *Self-Assessment.*

Launches typically involve a full class discussion of a problem situation and related questions to think about at the beginning of the lesson. In the print version, typically the teacher reads a few paragraphs about a problem and then leads a full class discussion of these questions. It is effective from time to time to vary the launch in the print version. I have found the use of video to be helpful because it is both captivating and informative. I have also found it very challenging and time consuming to find appropriate video that supports a launch.

A real benefit of the digital version is that it has many launches with embedded video and audio clips. Because it is embedded in the digital curriculum, it flows naturally and does not feel like an add-on. Not only does the digital version have video launches at the beginning of each lesson, but also for the second or third investigations. The use of video and audio engages students by allowing variation in the launch, and having it built into the material helps the teacher with time management. To compare the different versions, Figure 10.4 is the beginning of an investigation from the print material. The italicized question lets the student know what to look for in the investigation, but no launch video or audio is provided in

> **Investigation 2** How Surveys Can Go Wrong
>
> When interpreting survey results, it is important to consider how the survey was constructed and carried out. In this investigation, you will examine how bias may occur in a survey.
>
> As you work on the problems in this investigation, look for answers to this question:
>
> *What are the characteristics of an untrustworthy survey?*

Figure 10.4 Example opener of a student investigation found in the print material.

Figure 10.5 Example opener of a student investigation found in the digital material.

the print medium. In contrast, Figure 10.5 is the beginning of the same investigation in the digital version. They look similar but, as you can see, there is a built-in link (below the picture) where the student can view a very entertaining and relevant video. The video makes it clear how questioning patterns can lead to bias in surveys. The students found videos such as this one extremely engaging, entertaining, and motivating for launching an investigation about how surveys can go wrong.

GEAR 4: EXPLORATION OF PROBLEMS AND STUDENT GROUP USE OF GOOGLE DOCS

The most challenging standard of mathematical practice from the Common Core State Standards (National Governors Association Center for Best Practices & Council of Chief State School Officers, 2010) to implement in my classroom is: *Construct viable arguments and critique the thinking of others.* This is vital to student success as they explore problems. Even though my students work and discuss problems in groups when using a print curriculum, they ultimately write down their own ideas in their own way. Often students are on different parts of a problem because they are not communicating in an appropriate fashion. Students may settle to just have something written, even though a concept is not entirely clear to them. Frequently, technology is used in exploration. When using pencil and paper to record ideas, it is difficult to incorporate the technology piece in an explanation. Many of these issues disappear when using the digital materials. In part, this is because the digital format incorporates use of Google Docs. Each group has their own document that they share with each other and with the teacher. The group agrees on one answer, which is recorded in the Google Doc. When a group of students discuss a problem but need to record one answer that satisfies everyone, viable arguments and critiquing happen naturally. When one recorded answer represents the whole group, discussions are richer, and the students question each other and even argue about the mathematics much more frequently when using a Google Doc.

    An example problem and related student work are shown in Figures 10.6 and 10.7, respectively. In this example, the students took a screen shot of the first two targets, imported them into their document, and labeled them correctly. They then created their own targets using the Google Doc. I remember walking over to AJ during this lesson and whispering that we should have had targets drawn in the shell we provided for them for the investigation. AJ's response was, "Let's see what happens." No sooner had our conversation ended than I noticed the screen shot in Figure 10.7. It was very exciting for me to see how students were able to solve this problem creatively using technology. I was so excited that I grabbed the student's computer and plugged it into the LCD projector for the whole class to see. Importing into a Google Doc is something students naturally do, as noted earlier. Sometimes students do work on their calculators. What students

① As most people realize, the estimate from a sample probably will not be equal to the population parameter. Two possible sources of error are chance and bias.

- **Chance (or sampling) error** results from the fact that a survey based on a random sample does not ask everyone in the population. Thus, the estimate from the sample may not be exactly equal to the population parameter. In Lesson 3, you will learn that a larger sample size tends to reduce sampling error.

- **Bias**, on the other hand, tends to push the estimate to one side of the population parameter. Specifically, in repeated sampling, the estimate from the sample is too big or too small, on average.

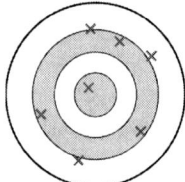 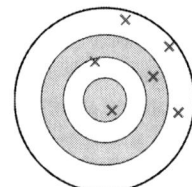

In the diagrams above, the center of the target represents the population parameter. The ×s represent estimates from the sample, attempts to hit the center of the target. There are four targets that show the different possibilities of error due to chance and bias:

- Error due to both chance and bias
- Errors due to chance alone
- Error due to bias, with very little chance error
- Almost no error due to bias or to chance

Identify the cases of error due to chance and bias that are represented by the given targets. Draw targets that show the remaining cases.

Figure 10.6 Example student task found in the digital materials.

will often do then is take a picture of their calculator screen and import that into their Google Doc. Because the Google Doc was so effective, I use a Google Doc in my classes today. It was the one piece that transferred from the Deeply Digital format to the Core-Plus Mathematics and Transition to College Mathematics and Statistics print classrooms. I look forward to the day when I will be teaching all of my classes in the Deeply Digital format.

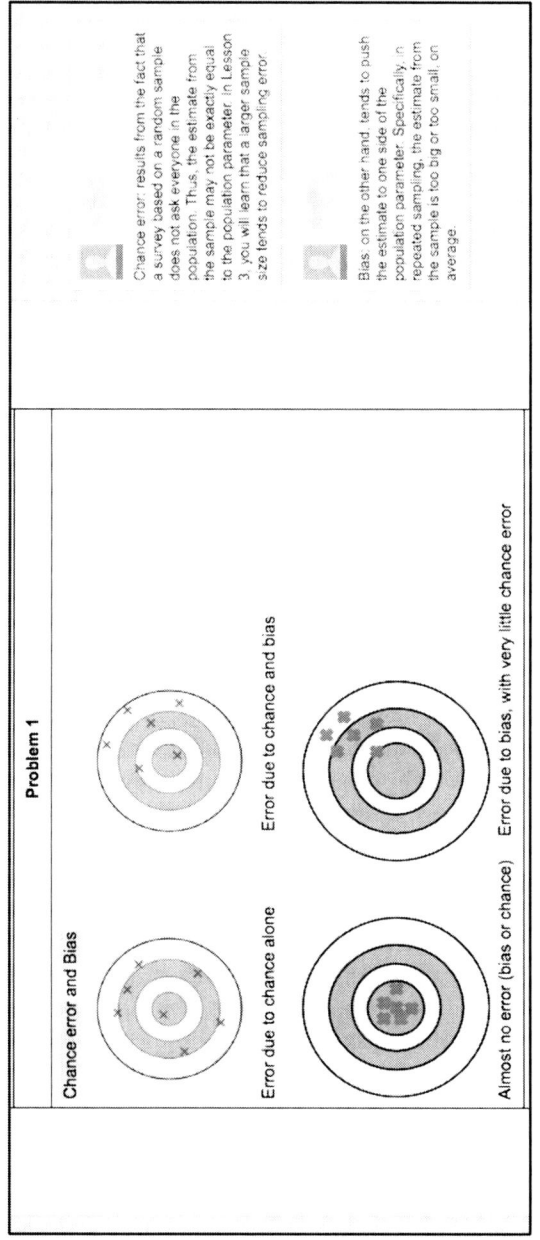

Figure 10.7 Example work from a group of students.

GEAR 5: EXPLORATION AND WHOLE-CLASS USE OF GOOGLE DOCS

The digital format of the instructional materials helped me with time management in my classroom in a couple of ways. One way it helped was in monitoring each group's progress. All groups work at different rates, and some think in different ways than others. When students are engaged to think deeply with open-ended questions, time management can be difficult. The digital materials afforded me the opportunity to easily monitor each group's progress and intervene if necessary. Because there was one Google Doc for each group, I needed only to monitor five different group documents. While using the print version, each student recorded his own answers and it was difficult to get a feel for where my students were.

A second way the digital materials helped with time management is they made certain classroom activities very efficient. There are some activities in investigations that are time consuming. Although the teacher may deem the activity worthwhile, the teacher may also consider skipping an activity due to the amount of class time it takes. These educational decisions are challenging for the teacher to make. It is easy to resort to "telling" my students what they need to know because of time constraints. This, I believe, detracts from student learning and understanding because students need time to experience the mathematics, react to these experiences, and ultimately draw conclusions. Here is an example of what I mean. Figure 10.8 is one of my favorite problems because the result was not what I expected.

As the problem states, it would be a lot of work to compute and average the areas of all 115 circles. So, a sample should be taken. If you think about this problem from a teaching perspective, the suggestion to sample is a good idea for the students. It will help speed up the process of estimating an average area. The suggestion to sample, however, does not help the teacher in Part c of the problem where you need to pool two lists of data. Each list will contain two average areas per student. My class had 20 students, so the students and I had to efficiently collect and manipulate 40 data entries. Using the print version, a good teacher would set up a spreadsheet before class and then enter accurately all 40 data points as the students report their results. This would take some time to complete even with a pre-made spreadsheet, and students lose focus in the process. The challenging question presented to the teacher becomes, "Is it worth it?" The temptation here is to simply read to the students the results that the teacher's edition supplies. This, however, would abuse the exploration step of the TCMS instructional model. The goal of the investigation is for students to learn that sample selection is an important part of conducting a trustworthy survey. There is no better way to obtain this goal than for students to experience this first hand. Time management is a daily tension for teachers.

Examine the set of circles shown on the following page. Your task in this problem is to estimate the average area of all 115 circles. It certainly would be a lot of work to compute the area of each circle, so this is a situation where sampling might be better. You will be comparing two different methods to get your sample.

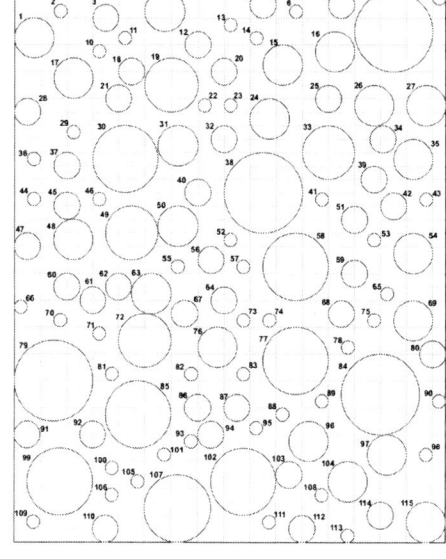

a. First use a **judgment sample**, using your best judgment to select five circles for the sample. That is, select five circles you think are fairly typical. Record the radius and area of each of your five circles. Finally, compute the average of the areas of the five circles.

b. Next, use technology to generate five random numbers between 1 and 115 inclusive and locate the corresponding circles. Find the average area of the five randomly selected circles.

c. Now, pool your data with other students in your class, making two lists and two number line plots for the averages generated in Parts a and b. Compare means of the two lists. Are the values about equally spread out in the two plots?

Figure 10.8 Example investigation problem from the digital materials.

The digital materials support teachers by providing options as they deal with time issues. Figure 10.9 is a spreadsheet that I used with students from my TCMS classroom. This spreadsheet was set up prior to class and reflects that students were arranged by their groups. There were specific cells marked for each student to enter his data. Because this spreadsheet was shared with the entire class, each student could observe and edit work from the entire class at the same time. As the students entered their data, they could see the average (mean) area calculated automatically and change as data were entered. The mean of the areas from the judgment samples were quite a bit higher than the mean from the random samples. The points of the problem that come out very clearly are that random sampling is a good method for sample selection and judgment sampling is a type of sample selection bias. The revision history on the right hand side of the spreadsheet shows that all of this activity was accomplished in just a few minutes. As you

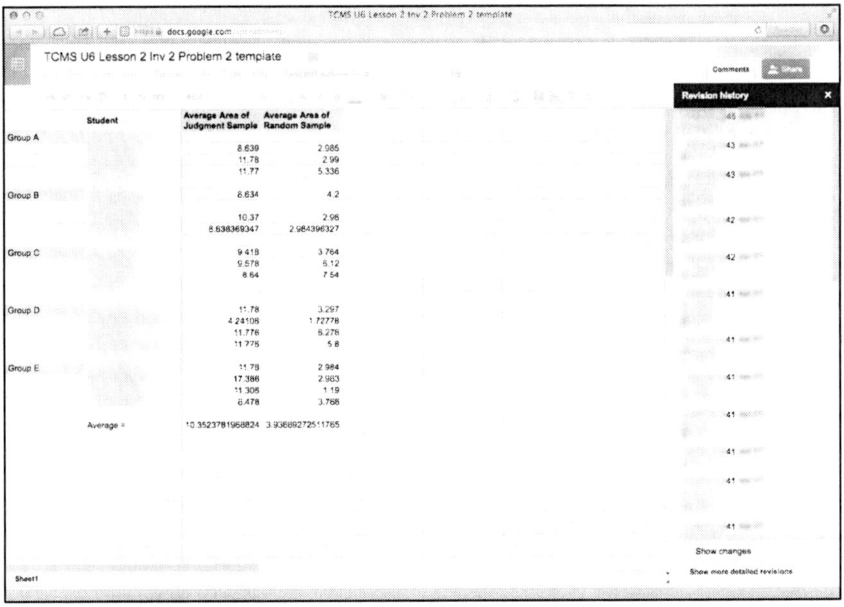

	Student	Average Area of Judgment Sample	Average Area of Random Sample
Group A			
		8.639	2.985
		11.78	2.99
		11.77	5.336
Group B		8.634	4.2
		10.37	2.96
		8.638369347	2.984396327
Group C		9.418	3.764
		9.578	6.12
		8.64	7.54
Group D		11.78	3.297
		4.24106	1.72778
		11.776	6.278
		11.775	5.8
Group E		11.78	2.984
		17.386	2.963
		11.306	1.19
		8.478	3.768
	Average =	10.3523781968824	3.936889272511765

Figure 10.9 Sample student work using spreadsheets.

might imagine, this was an engaging activity for my students. They enjoyed representing themselves by supplying their own data and seeing how their work contributed to the problem solution. Because it was so quick, students did not lose focus. They got it done and the results for the average areas for both methods of selection were instant. This allowed the students and the teacher to have an immediate summarizing conversation about bias in sampling and how important a random sample actually is. Using spreadsheets allowed students to explore and experience the statistics as the instructional model intends. It also made the exploration happen in very little time.

GEAR 6: SHARE AND SUMMARIZE USING STUDENT WORK

The Share and Summarize phase is an important step in the instructional model. After students explore different problems focusing on important ideas, students shift their focus to answering discussion questions such as those found in Figure 10.10. Whole-class discussion of key concepts and methods developed by different groups leads to a constructed summary of the important mathematical ideas.

Summarize The Mathematics

> **Summarize The Mathematics**
>
> In this investigation, you learned that sample selection is an important part of conducting a trustworthy survey. Equally important is getting a good response from each person selected for the sample.
>
> (a) Explain why random sampling is a good method for sample selection.
>
> (b) What are some types of sample selection bias?
>
> (c) What are some types of response bias?
>
> *Be prepared to share your ideas and examples with the class.*

Figure 10.10 Example Summarize the Mathematics section from the digital materials.

When using the print version of the book, the teacher reads the questions and the students provide their responses. Students are encouraged to write answers so they make sense to them. Sometimes a teacher may decide it is necessary to share and summarize before the end of an investigation. Figure 10.11 shows an investigation problem from the digital version of the instructional materials of TCMS.

Enabling each classroom group to share its Google Doc with the teacher allows for a different level of interaction between the students and their teacher. The Google Doc allows the teacher immediate access to their work. Sometimes this would lead to immediate intervention. Figure 10.12 shows the sample student work that groups produced for the problem shown in Figure 10.11. The different colors refer to the different responses from the various groups in the class.

It is easy to see which are correct and which need improvement. Although this is important, we can all learn by examining different responses. I think too often what is right and what is wrong becomes our focus as teachers. Better questions may include: What do they all have in common? How are they different? What can we gain from looking at different interpretations of the same question? I showed the work displayed in Figure 10.12 to my students during the Share and Summarize and it provided a springboard to an interesting discussion about probability distributions. Students understood the concept better because we looked at a variety of interpretations of what they created, how they created it, and the features that make up a probability distribution. By clicking on the glossary in the resource

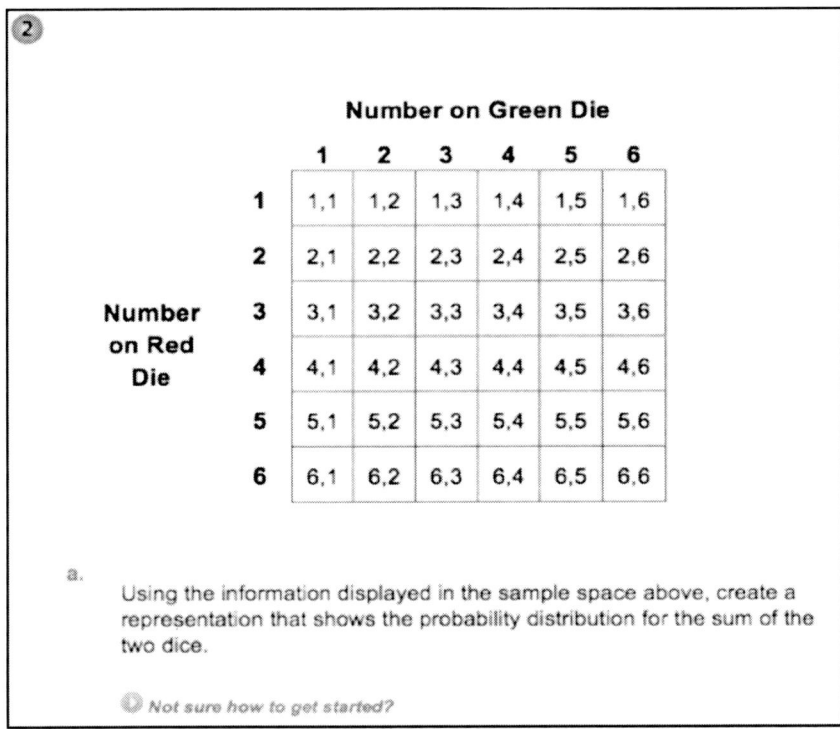

Figure 10.11 Example investigation problem from the digital materials.

navigation window (see Figure 10.13), a person can find information about a probability distribution.

Many of the groups represented their probability distributions using tables. Students pointed out that the "red" and "orange" answers were similar. I asked, "What could make them better?" and directed them to the "green" table. They noticed how important the labels were in the "green" table, and we settled on "sum" and "probability" for the two labels. The "green" group was quick to notice they had forgotten the denominator in their frequency column. Having the students realize this on their own is much more effective then someone telling them of their mistake. A summarizing idea that was fleshed out because of this mistake was *the sum of the numbers in the probability column should always equal one*. Students noticed the "blue" response was the most different. I used the "blue" response to get at the graphical form that the glossary definition mentions. As shown in Figure 10.14, I rotated their graphs and drew the blue vertical lines. Although we didn't fully complete this graphical form, we talked about both the label and scale of the *y*-axis. It was helpful for the students to see their

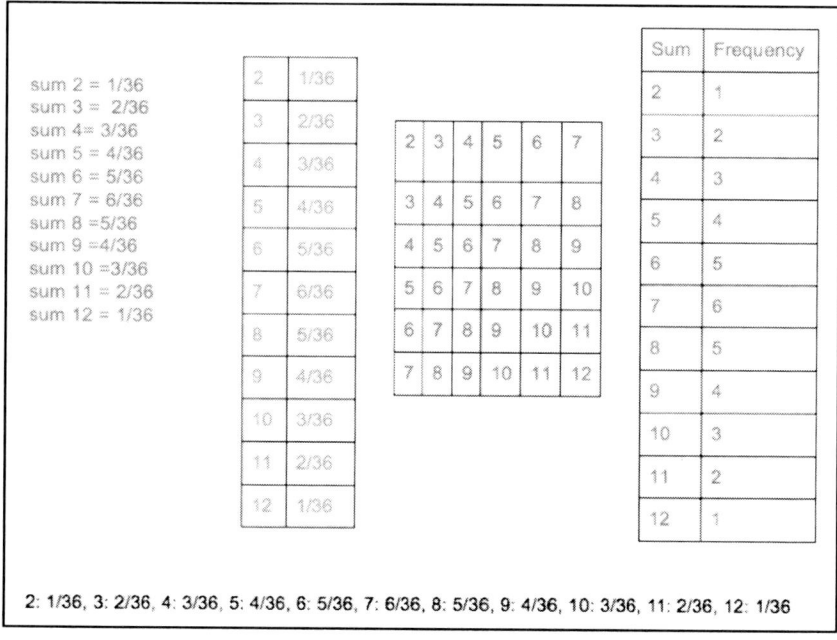

Figure 10.12 Sample student work for an investigation problem.

Probability Distribution: A description of all possible quantitative (numerical) outcomes of a chance situation, along with the probability of each outcome; the distribution may be in table, formula, or graphical form.

Figure 10.13 Example definition in the glossary of the digital materials.

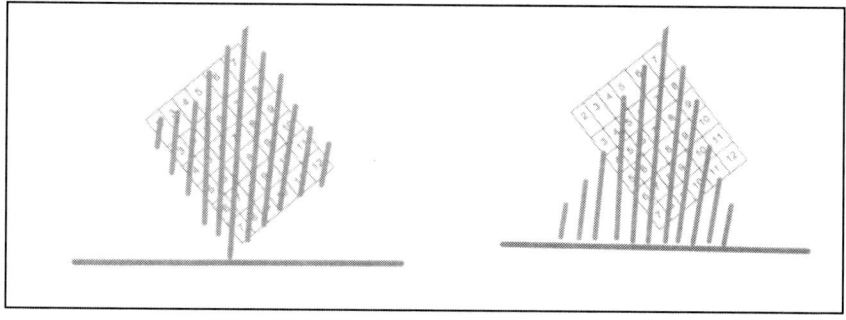

Figure 10.14 Example: Moving towards the graphical representation of a probability distribution.

work compared to other groups' work. This process of comparison is an effective way to summarize the mathematics. I have tried this process of comparison before, when using the print form of the book. It is as effective but takes longer to pull off. There are the added steps of making a poster of their work so it is viewable by all. Then it needs to be posted for all to see. Because of the time factor, it is not always feasible to use this method in the print version.

THE REVERSE GEAR: LOOKING BACK ON THE USE OF THE DIGITAL MATERIALS

Being part of this work has been extremely rewarding. It once again makes me think hard about how students learn and what we as educators can do to continue our efforts of meeting students where they are. Technology affords many ways for the teacher to reach students. The digital shell is elegant, functional, and engaging. It caused my students to think deeply. Asking the "hard" questions first and then supplying the scaffolding later gave my students choice of when or if to seek help. Students were engaged in the launch through the use of video and audio, and the video and audio served the teacher well, too, saving the teacher time and energy. Using the materials, the students created viable arguments and critiqued each other. The use of the Google Doc during the exploration time allowed me to efficiently monitor their work and the pace of each group, which at times would lead to immediate intervention. The Google Doc also provided a means to have a whole classroom discussion around student thinking while summarizing the mathematics, which is very powerful.

The deeply digital curriculum is the merging of reform mathematics curriculum and technology. It has greatly influenced my philosophy of education. The digital materials discussed earlier influenced the way the students interacted with the mathematics in a positive way. The digital curriculum gives students the keys and puts them in the driver's seat. Being part of this experience has been exciting for both the teacher and the students. We were thrilled to take this digital material for a test drive and look forward to driving this Lamborghini in the future!

ACKNOWLEDGMENTS

I want to thank Dr. AJ Edson for his expertise in the design and development of these digital materials and for the opportunity given to my students and me to use these materials. It was both thought provoking and rewarding for both students and teacher. It has reinvigorated my teaching, and

I look forward to using digital materials in the future. I would also like to acknowledge the work of James Laser, Western Michigan University, in coding the initial shell for the digital unit. The photos from the Transition to College Mathematics and Statistics project are presented here with permission from the project.

REFERENCES

Edson, A. J. (2016). A design experiment of a deeply digital instructional unit and its impact in high school classrooms. In M. Bates & Z. Usiskin (Eds.), *Digital curricula in school mathematics* (pp. 177–193). Charlotte, NC: Information Age.

National Governors Association Center for Best Practices & Council of Chief State School Officers. (2010). *Common Core State Standards for Mathematics.* Washington, DC: Authors.

PART III

RESEARCHING DIGITAL CURRICULUM

CHAPTER 11

ANALYSIS OF EIGHT DIGITAL CURRICULUM PROGRAMS

Jeffrey Choppin[1]

In this chapter, I build on previous work in which my research team and I developed a typology for analyzing digital curriculum programs (Choppin, Carson, Borys, Cerosaletti, & Gillis, 2014) by expanding the number of analyzed programs to eight, by providing more details of our analysis, and by touching on themes from the CMSC Third International Curriculum Conference held in Chicago in November 2014 (hereafter called the CMSC conference). In the typology, we explored the potentially transformative characteristics of digital curriculum programs described in the literature (Abell, 2006; Fletcher, Scaffhauser, & Levin, 2012; Selwyn, 2007; Zhao, Zhang, & Lai, 2010), such as the use of multimedia, greater interactivity, increased and more varied social interactions, greater individualization and customization, and assessment systems embedded into the programs that allow for rapid and visual reporting of student results.

Two trends make the analysis of the current state of digital programs in the United States particularly relevant. First, there is an increasing policy push for teachers to use digital materials in Korea, the U.K., the United States, and elsewhere (Devaney, 2013; LEAD Commission, 2012; Selywn, 2007; Usdan &

Gottheimer, 2012). Second, two-thirds of U.S. middle school mathematics teachers report using curriculum materials from digital or electronic sources at least once a week, and many more often than that (Davis, Choppin, Roth McDuffie, & Drake, 2013), showing that digital resources in middle school mathematics classrooms have become a staple of classroom life.

BRIEF SUMMARY OF TRANSFORMATIVE FEATURES OF DIGITAL MATERIALS

Digital curriculum materials can incorporate multimedia, which include high-definition graphics, video clips, animations, simulations, interactive lessons, and virtual labs (Fletcher et al., 2012), though Mayer (2003) states that the coordination of multimedia components—not simply their presence—influences their impact on learning. Interactivity in digital materials includes: non-linear media such as hyper-linked texts; incorporation of web-based resources to design lessons; models or tools that allow the user to see how change in parameters or sliders alter a scenario or context (Dede, 2000); and, in light of the CMSC conference, tactile interfaces in which the user's finger motions directly manipulate objects in a micro-world environment (Laborde, 2016; Sinclair, 2016). Programs that are connected to the Internet provide the opportunity for new kinds of virtual interactions with people, including those with particular expertise, across time and space (Anderson, 2005; Dalsgaard, 2006; Zhao et al., 2010). One of the most touted features of digital programs is their potential to customize content based on a person's learning needs and styles. This is often accomplished via embedded assessment systems that automatically store, analyze, and transmit student performance and activity completion data (Abell, 2006; Fletcher et al., 2012; Ready, 2014).

FEATURES DISCUSSED AT THE CMSC CONFERENCE

In addition to the features discussed above, additional features emerged in the presentations at the CMSC conference. One feature prominent in the SMART Korean textbook (Lew, 2016) and the VisualMath digital textbook (Yerushalmy, 2016) was the *ubiquitous presence of tools* for students to use as they work on problems. These tools may be tailored to particular topics, as is the case with VisualMath, or constant across topics, as is the case with the Korean textbook. The ubiquitous use of tools provides forms of interactivity not possible in paper textbooks and provides task customization and convenient access. These tools contrast with the interactive diagrams described by Dede (2000) in that their use is more open-ended and allows greater opportunities for students to explore and to author approaches to problems.

A second feature that emerged at the conference is a refinement of the notion of *customization* or *differentiation*. Scholars, including Confrey (2016), make an important distinction with respect to customization. A goal of customization or differentiation of curriculum programs should be to strive to "achieve a balance between characteristics of the learner and those of the learning environment, between what is challenging and productive and what is beyond the student's present capabilities" (Keefe, 2007, p. 221). This can be done via individualization or customization, where individualization is seen as a learner engaging almost exclusively with a digital curriculum program, with little interaction with others around similar academic tasks. Although such a program may appeal to needs specific to that individual, it removes a key feature of personalization, which requires "interactive learning environments designed to foster collaboration and reflective conversation" (Keefe, 2007, p. 221). Consequently, personalization involves embedding scaffolds for an individual student within a commonly shared task, so that students can access them only as needed. This will help all students have opportunities to engage in tasks in the most cognitively demanding way.

A third feature is focused more on logistical features of using a digital curriculum program, more broadly termed *interoperability*. Interoperability involves the adaptation of the program across platforms, the need for consistent Internet access, and the bandwidth necessary to operate a program productively.

A fourth feature involves the *formative versus summative* nature of the display of assessment data, which Phil Daro encapsulated in his distinction between dashboards and windshields. What teachers want is windshields, he argues, that provide useful assessment data accessible to teachers in real time while being minimally distracting.

METHODS

Sampling

We focused our analysis on digital curriculum materials that can supplement or replace conventional paper textbooks through the use of computers, electronic tablets, or similar devices that allow for one-to-one access. Most, if not all, of the resources analyzed have an online component or are entirely web-based. The digital curriculum platforms we analyzed include those from major publishers and vendors, those that have received media attention, and those that we identified as having some unique characteristics that merited attention. A goal of our sampling was to capture a range of digital programs available in the United States.

Types of Programs Analzyed

We analyzed eight programs, including those we categorized as individualized learning programs, collections of digital lessons, digitized versions of traditional textbooks, and a microworld environment designed to be a comprehensive algebra curriculum.

The individualized learning programs included:

- ALEKS, a web-based adaptive program that provides explanations and practice of skills and procedures;
- Dreambox, an adaptive program that provides individualized mathematics instruction through interesting contexts and interactive manipulatives; and
- Khan Academy, a web-based program designed to have students work at their own pace through videos of narrated presentations on concepts and procedures, online practice problems, and online assessments.

The collections of digital lessons included:

- LearnZillion, a collection of recorded presentations developed by a team of over 100 teachers; and
- YourTeacher (MathHelp.com), a collection of video narrations of problems and worked out solutions.

The digitized versions of traditional textbooks included:

- *CINCH*, a version of a traditional textbook program designed to be used strictly in a digital environment and not accompanied by a paper textbook, though most of its content can be downloaded as PDFs; and
- ConnectED (Glencoe Math), a platform offering digital access to the full range of grade-level content and resources for students and teachers for certain McGraw Hill textbook series, such as *Glencoe Math Courses 1, 2,* and *3,* and the 4th edition of *Everyday Mathematics*. The first of these series was analyzed.

The microworld environment we analyzed was *Algebra in Action*, a digital program written for the iPad that weaves together an Algebra curriculum and a narrative.

Analysis

In our analysis, we looked at three major categories:

1. How students might interact with content in the programs;
2. How teachers could adapt or use (re-source) content; and
3. How the programs incorporated built-in assessment systems.

For each program, we sampled content until additional sampling revealed trends already observed in the programs, usually around 15 to 20 lessons or topics.

To look at features of how students might interact with the programs, we analyzed students' learning experiences within the program and the features related to individualization or customization. For students' learning experiences, we considered what kinds of activities students would be expected to do while engaged with the materials, the level of interactivity, and the ways content was introduced and presented to students. In terms of how teachers could use or adapt the materials, we considered how teachers might be able to revise or adapt the materials to suit their goals and contexts. This included the ability to select and revise program content for use inside or outside of the program. To analyze how the programs incorporated built-in assessment systems, we explored how student performance was elicited, recorded, evaluated, and displayed within the program, and how that performance influenced choices for content selection and pacing, whether done automatically (adaptively) or controlled by the teacher. The display of data was almost exclusively provided from a dashboard—versus windshield—perspective in that the data did not relate to real-time student performance in a given lesson. The dashboards provided quick access to assessment data, but typically this was student performance after the fact to inform future instructional decisions. A windshield-type system, by contrast, would provide a real-time data feed of students' work (individual and collective) that would inform in-the-moment decision making and be easily accessible as the teacher monitored the class.

RESULTS

Student Learning Experiences Within the Programs

The results for how students would interact with programs are presented in Table 11.1, in which the programs are ordered according to the categories previously listed. Key findings in the table are discussed below.

TABLE 11.1 Features That Affect Students' Interactions With the Programs

Name of Program	Learning Experiences	Differentiation/Individualization	Social/Collective Features
ALEKS	Students take assessments whose results influence which types of problems they are assigned within the program. Students work on problem sets created by the program; once they have shown mastery, they move on to new types of problems. Students can view explanations of content if they are unable to solve the practice problems.	Students may be assigned problem sets by the teacher or they may use their personalized pie chart to select problem areas identified by the program's assessments as areas in which students need additional practice.	Teachers and students have the ability to communicate with each other within the program through an internal mailbox system and a Class Forum discussion board feature available on the homepage for each student.
Dreambox	Students are given a choice of units to explore and encouraged to "play" until they receive a gold coin. Students are given tasks that have multiple possible solutions and are provided with virtual manipulatives to help them solve the problems. Hints are provided if the students need them.	The program performs ongoing assessments to determine the next lesson for the student. The program evaluates the students' use of manipulatives to determine which hints to provide and to select subsequent tasks.	The program does not have discussion boards, wikis, or other ways within the program for teachers and students to communicate.
Khan Academy	The student is presented with a series of video topics followed by practice problems related to the skill presented in the video segment. The students enter their solutions to the problems in blank windows. If their answers are incorrect, the program gives the students suggestions or hints for solving the problems.	The program is designed for independent work and is individualized by student performance on practice problems that follow each video. The program creates more practice problems or moves students onto new content depending on student success on practice problems.	In the questions area, a student or coach is able to ask or answer any question. This is asynchronous communication (it can take 20 minutes or longer for a response), and it is also unclear whether another student or a coach is answering the question.

(continued)

TABLE 11.1 Features That Affect Students' Interactions With the Programs (continued)

Name of Program	Learning Experiences	Differentiation/Individualization	Social/Collective Features
LearnZillion	Teachers can use the videos provided on the website as a whole class presentation or assign them to individual students. Each video focuses on a specific standard from the Common Core State Standards for Mathematics (CCSSM; National Governors Association Center for Best Practices & Council of Chief State School Officers, 2010). Each lesson has a multiple-choice quiz and guided practice section. The guided practice includes a video of a sample problem. Students can pause the video, work out the solution, then play the rest of the video.	Teachers have the opportunity to assign lessons to the entire class or individual students. This allows for the teacher to customize learning for each student. Once a lesson is assigned and students gain access to it (on a certain date specified by the teacher) they may work through all of the lessons and quizzes available to them.	There is no within-site means to communicate.
Your Teacher (MathHelp.com)	Each lesson consists of example problems with a video or narration of a teacher explaining and working out a problem. There are a number of practice problem options which can only be solved offline and checked online by clicking on the checkmark button. A hint button provides a tip and the WORK button shows all of the work for the problem. A "deep thought problem" provides a challenge question. This problem is accompanied by narrated solution.	The site tracks scores on self-tests and suggests areas of weakness for scores lower than 80%. Students scoring below 80% are directed to lessons for that content and must continue to retake the self-test (which changes for each iteration) until they score above 80%. The program provides individualized grade reporting and progress tracking.	There is no within-site means to communicate.

(continued)

TABLE 11.1   Features That Affect Students' Interactions With the Programs (continued)

Name of Program	Learning Experiences	Differentiation/Individualization	Social/Collective Features
CINCH	Material is presented in a context, a procedure is demonstrated, and then students practice the procedure. In inquiry labs, students follow a set of predetermined steps as they answer short-response items.	The teacher can select assignments for particular students and provide tips or advice to students in the program.	There is an internal discussion board integrated with google docs. There are also personalized notes, highlight sections, and a chat window.
ConnectED (Glencoe Math)	Students can read the eBook, watch presentations, complete assessments, utilize virtual manipulatives, or work on teacher-generated material, worksheets, and other resources.	Lesson plans provide lists of differentiated homework assignments and instructional tips to teachers for remediation and enrichment options.	Teachers can send messages to students. Students cannot send messages or respond to a teacher's message.
Algebra in Action	The curriculum involves a narrative in which students fight invading aliens. Students watch videos of concepts or problems or play interactive games. There are embedded assessments in which student responses are saved for teacher evaluation.	The teacher determines the pace of the lessons. Teachers can determine if individual students need extra help with a concept based on the results of embedded assessments scored by the teacher.	Teachers can communicate with students via comments on problem sets, quizzes, and labs. There are no other features for communicating with other classmates or the instructor.

The results show that two of the three individualized learning programs (ALEKS and Khan Academy) focus on explanation and mastery of mathematical procedures, while the third (Dreambox) provides a more interactive and conceptual experience for students. In Dreambox, for example, students "snap" numbers into bars to create equivalent number sentences, in the process decomposing larger numbers into smaller ones. This activity is designed for students to make sense of number composition and equivalence without an accompanying explanation or demonstration of an established algorithm. Two of these programs (ALEKS and Dreambox) are adaptive, in the sense that assessments embedded within the program determine how content is sequenced for individual students.

The two collections of digital lessons, LearnZillion and YourTeacher, provide sets of videotaped explanations or demonstrations for students to watch, with no opportunity for interactivity. In LearnZillion, teachers may assign topics to students based on performance on embedded assessments; YourTeacher makes suggestions for content based on student test scores.

In the two digitized versions of traditional textbooks, CINCH and ConnectED, the student interaction with the program is similar to that of a traditional textbook, with the exception that students have access to multi-media content such as presentations or explanations and have access to some basic manipulatives or measuring tools. Similar to Dreambox, Algebra in Action engages students in interactive simulations or activities that help students explore content without having to apply established algorithms. Like some of the other programs, the content can be individualized via teacher monitoring of student performance within the program as well as the potential to provide feedback to students within the program.

Teacher Use and Adaptation of Programs

Table 11.2 provides a summary of how teachers can draw from and adapt program content in flexible ways to suit their goals and contexts. None of the individualized learning programs nor Algebra in Action offers teachers the opportunity to revise content or tools within the program. Two of these programs, ALEKS and Dreambox, provide the possibility for the program to automatically select and sequence content based on student performance on the embedded assessments, while in Khan Academy teachers may recommend but cannot limit content students are able to access.

In the two collections of digital lessons, LearnZillion and YourTeacher, the only possibility to revise content was for teachers to modify the PowerPoint files that accompany the videos in LearnZillion. In the two digitized versions of traditional textbooks, CINCH and ConnectED, teachers are able to select video presentations and activities to insert into the lesson

TABLE 11.2 Features for Curriculum Use and Adaptation

Name of Program	Ability to Map and Sequence Lessons	Ability to Design Content of Lessons	Ability to Locate and Use Multi-Media Presentation Materials
ALEKS	The program uses a teacher-determined pacing chart to determine the content that can be accessed by students and creates a unique program by evaluating student performance on practice problems. The program generates practice problems in response to student performance.	The teacher assigns a worksheet that she created, or the program gives a worksheet based on student performance. The teacher can upload instructor notes, class notes, and outside resources into the lessons.	There are no multi-media resources within the program that the teacher can use to create or customize lessons.
Dreambox	The program suggests the sequencing of content based on student performance, though students may select content from topics deemed within their level of competency.	Teachers have access to virtual manipulatives for large group instruction on an interactive whiteboard.	There are no multi-media resources within the program that the teacher can use to create or customize lessons.
Khan Academy	The teacher may select a specific topic such as writing expressions or finding absolute values from the knowledge map. There is no mechanism to set up a specific sequence of lessons, and students are able to access any video or exercise at any time.	There is no ability for teachers to customize any videos or exercises.	Teachers may assign videos to students in which a topic or procedure is explained. Videos are searchable by topic.
LearnZillion	The teacher can create links to specific lessons and save them to a calendar for planning and sequencing purposes. The content does not appear on the students' dashboard until the teacher assigns it.	Teachers do not have the ability to change any video or quiz content but do have the ability to download the slides used in the video as a PowerPoint and customize them as needed.	There are PowerPoint slides available for teachers to use to customize lessons. The files are searchable by CCSSM standard.

(continued)

TABLE 11.2 Features for Curriculum Use and Adaptation (continued)

Name of Program	Ability to Map and Sequence Lessons	Ability to Design Content of Lessons	Ability to Locate and Use Multi-Media Presentation Materials
YourTeacher	Lessons can be browsed by a list or searched by textbook or keyword. The software sequences content based on a manually selected topic or a textbook. Mapping can be done manually by selecting topics in a predetermined order. However, students can access any topic at any time from the menu.	There is no facility for teachers to upload their own notes or outside resources or to customize materials. The assessments are all built into the site and are not modifiable.	Teachers may assign or show videos to students in which a topic or procedure is explained.
CINCH	The teacher has options for creating a course schedule that includes days of school/instruction and lesson sequence. Previously taught classes are stored on the database and can be accessed in the future.	The program has a stock of lessons that makes up the core of its curriculum but allows for the addition and customization of presentation material and support activities/videos/online resources that get added and saved to a lesson.	There are applets, videos, and virtual tools available for teachers to use to create or customize lessons.
ConnectED (Glencoe Math)	The program provides lesson plans based on textbook selection. Teachers can access a calendar-based planner to determine the lessons that can be accessed.	The lesson plans can be modified by adding new teacher supplied content and removing provided content.	There are applets, videos, and virtual tools available for teachers to use to create or customize lessons.
Algebra in Action	Teachers do not have the ability to alter the sequence of lessons.	The lessons, activities, quizzes, explorations, videos, problem sets, and quizzes cannot be altered by the teacher.	There are no multi-media resources within the program that the teacher can use to create or customize lessons.

plans they can create within the programs and to import external resources into these lesson plans.

Assessment Systems

The results in Table 11.3 show that all of the programs, with the exception of Algebra in Action, can display dashboard summaries of individual or whole class results on performances, with reports showing progress on particular standards. These displays are of the dashboard variety, showing summative data on the percent of skill-based items successfully completed. Two programs, Khan Academy and LearnZillion, show viewing data, such as which videos were viewed by participants or how long participants were logged into the program. Dreambox reports students' use of manipulatives including descriptions of strategies.

Teachers have no ability to alter assessments in the individualized learning programs, the collections of lessons, or Algebra in Action; in ALEKS and Khan Academy, teachers can suggest or assign certain assessments. In the digitized versions of traditional textbooks, CINCH and ConnectED, teachers can generate assessments by selecting items from databanks of existing items.

DISCUSSION

In the typology study (Choppin et al., 2014), we discussed how most of the programs we analyzed, except Algebra in Action and Dreambox, lacked transformative uses of multi-media or interactivity, such as characteristics that were evident in several of the programs presented at the CSMC conference, including VisualMath, the Korean SMART textbook, Cabri, and Amplify. Most of the programs we analyzed used multimedia to present recorded demonstrations of content rather than to create interactive tools or scenarios, were minimally adaptable, and based their assessment systems largely on reporting percentage rates of accurate completion of low-level tasks. Furthermore, with the exception of Algebra in Action and Dreambox, the view of mathematics embedded in these programs was based on procedural accuracy and mastery, with very limited presence of rich problems and opportunities for productive whole class discussions. By contrast, VisualMath, the Korean SMART textbook, and Amplify included problems that were accessible, could be approached using a variety of methods, and incorporated interactive tools that were ubiquitous and easy to access.

Embedded assessment systems in digital programs offer potentially powerful tools to help teachers gain a rapid understanding of students' mathematical understanding and procedural fluency. However, the main assessment

TABLE 11.3   Analysis of Assessment Systems

Name of Program	Ability to Create Online Assessments	Ability to Record and Evaluate Results of Assessments	Ability to Generate Dashboard or Other Summaries of Data
ALEKS	The teacher can assign quizzes and assessments and can choose a topic, the number of questions, grading scale, and completion date of quizzes.	The program records all results of student performance on the assessments and uses those results to move a student to a new topic within the parameters of the teacher designated pacing guide.	The program displays pie charts segmented into different content domains or concepts, with parts shaded for the program's evaluation of what the student has learned. Other reports show progress the student is making on the CCSSM standards, how much time a student has spent in the program, and a summary of basic facts performance.
Dreambox	Teachers have no ability to create assessments within the program.	The program automatically records the progress of the student.	There are dashboards that provide information in several formats and at the level of individual or whole class. The dashboard reports in percentages how far a student has progressed with respect to a particular standard, how much time the student has spent on this standard, and how many lessons were needed to work through the concept.
Khan Academy	Teachers have no ability to create or modify assessments within the program. The teacher (or student) is able to set individual "goals" of specific skills that a student should accomplish.	The program stores data on the videos viewed by students, time spent watching videos, exercises a student has completed, the student's solution steps, and the hints accessed by the student.	The teacher can monitor the progress of the whole class or individual on a class dashboard. The dashboard reports specific skills or a more general state of progress on the whole program for the whole class or for an individual.

(continued)

TABLE 11.3 Analysis of Assessment Systems (continued)

Name of Program	Ability to Create Online Assessments	Ability to Record and Evaluate Results of Assessments	Ability to Generate Dashboard or Other Summaries of Data
LearnZillion	Teachers do not have the ability to customize assessments.	The program tracks each student's activity within the program by showing the proportion of the assigned work that has been completed. The program records students' responses in the quizzes but not the guided practice.	A teacher can see what videos were watched and how a student did on the assessments from the dashboard. The number of students who completed these assignments and their performance is displayed as a whole class and shows what topics need further instruction.
Your Teacher (MathHelp.com)	Teachers cannot create assessments within the program.	The results of the self-tests are recorded in the system. The site tracks scores on self-tests and suggests lessons to address weaknesses for individual self-test scores lower than 80%.	Progress is tracked by showing percentage correct for each lesson and test as well as number of completed tests. The percentage correct by lesson and number of completed lesson tests overall are shown in a report card view.
CINCH	The assessments can be generated from a database of questions. An offline assessment option allows the teacher to print out the questions.	For online and clicker assessments, the program records and stores students' scores and responses.	Teachers can select reports that show students' progress by summary (class averages), item analysis (statistics about each question), or individual student reports in a pop-up window.
ConnectED (Glencoe Math)	The teacher can select questions randomly from a database or create their own questions. These questions can be specified by question type, source, or difficulty level.	Student responses and scores for online assessments are stored. Teachers can view and comment on student responses.	Reports of individual and class performance are available. Performance can be compared between classes. The results are presented with graphics showing percent correct and proficiency level.
Algebra in Action	Assessments are embedded into the curriculum, but the teacher cannot revise them.	Responses to quizzes, labs, and problem sets are stored online so the teacher can access and comment on them. Some responses are automatically scored while others are teacher scored.	There is no dashboard or other summaries of data.

focus of most of the programs we analyzed was on diagnosis and remediation of students' computational and skill mastery. Although a focus on skill remediation can be desirable in contexts in which many learners are far below grade level computationally (Ready, 2014), such a focus lacks the potential to inform teachers of more nuanced aspects of their students' understanding and ability to operate on mathematical objects. Jere Confrey (2016) and Phil Daro reported at this CSMC conference that the curriculum programs they helped develop had more robust and nuanced assessment systems, including features that helped teachers attend in real time to student understanding and performance, but these features were not presented at the conference.

One of the key features that will influence the success of digital programs is interoperability, which involves the adaptation of a program across platforms, the extent to which a program works with existing school Internet protocols, and other logistical requirements necessary to operate a program productively. An extension of the results presented in this chapter would be to analyze the interoperability of digital curriculum programs and to study the impact of interoperability on how the programs get taken up by teachers and students in classrooms.

NOTE

1. This research was supported in part by the National Science Foundation under grant No. DRL-1222359. The opinions expressed herein are those of the author and do not necessarily reflect the views of the National Science Foundation. I would like to acknowledge Cynthia Carson, Zenon Borys, Cathleen Cerosaletti, and Rob Gillis for their work analyzing the programs. I would also like to acknowledge Jon Davis, Corey Drake, and Amy Roth McDuffie for providing feedback on earlier drafts of the typology manuscript.

REFERENCES

Abell, M. (2006). Individualizing learning using intelligent technology and universally designed curriculum. *Journal of Technology, Learning, and Assessment, 5*(3), 4–20.

Anderson, T. (2005). *Distance learning—Social software's killer app?* Paper presented at the Open and Distance Learning Association of Australia 2005 Conference. Retrieved from http://www.unisa.edu.au/odlaaconference/PPDF2s/13%20 odlaa%20-%20Anderson.pdf

Choppin, J., Carson, C., Borys, Z., Cerosaletti, C., & Gillis, R. (2014). A typology for analyzing digital curricula in mathematics education. *International Journal of Education in Mathematics, Science, and Technology, 2*(1), 11–25.

Confrey, J. (2016). Designing curriculum for digital middle grades mathematics: Personalized learning ecologies. In M. Bates & Z. Usiskin (Eds.), *Digital curricula in school mathematics* (pp. 7–33). Charlotte, NC: Information Age.

Dalsgaard, C. (2006). Social software: E-learning beyond learning management systems. *European Journal of Open, Distance and E-Learning*. Retrieved from http://www.eurodl.org/?article=228

Davis, J., Choppin, J., Roth McDuffie, A., & Drake, C. (2013). *Common Core State Standards for Mathematics: Middle school teachers' perceptions*. Rochester, NY: Warner Center for Professional Development and Education Reform.

Dede, C. (2000). Emerging influences of information technology on school curriculum. *Journal of Curriculum Studies, 32*(2), 281–303.

Devaney, L. (2012). Education chief wants textbooks to go digital. *eSchool news: Technology news for today's K–12 educator*. Retrieved from http://www.eschoolnews.com/2012/10/03/education-chief-wants-textbooks-to-go-digital/

Fletcher, G., Scaffhauser, D., & Levin, D. (2012). *Out of print: Reimagining the K–12 textbook in a digital age*. State Educational Technology Directors Association.

Keefe, J. W. (2007). What is personalization? *Phi Delta Kappan, 89*(3), 217–223.

Laborde, J.-M. (2016). Technology-enhanced teaching/learning at a new level with dynamic mathematics as implemented in the new Cabri. In M. Bates & Z. Usiskin (Eds.), *Digital curricula in school mathematics* (pp. 53–74). Charlotte, NC: Information Age.

LEAD Commission. (2012). *Leaders discuss transition to digital textbooks*. Leading Education by Advancing Digital. Retrieved from www.leadcommission.org.

Lew, H. (2016). Developing and implementing "smart" mathematics textbooks in Korea: Issues and challenges. In M. Bates & Z. Usiskin (Eds.), *Digital curricula in school mathematics* (pp. 35–51). Charlotte, NC: Information Age.

Mayer, R. (2003). The promise of multimedia learning: Using the same instructional design methods across different media. *Learning and Instruction, 13*, 125–139.

National Governors Association Center for Best Practices & Council of Chief State School Officers. (2010). *Common Core State Standards for Mathematics*. Washington, DC: Authors.

Ready, D. (2014). *Student mathematics performance in the first two years of teach to one: Math*. New York, NY: Teachers College, Columbia University.

Selwyn, N. (2007). Curriculum online? Exploring the political and commercial construction of the UK digital learning marketplace. *British Journal of Sociology of Education, 28*(2), 223–240.

Sinclair, N. (2016). New starting points for number sense using *TouchCounts*. In M. Bates & Z. Usiskin (Eds.), *Digital curricula in school mathematics* (pp. 205–221). Charlotte, NC: Information Age.

Usdan, J., & Gottheimer, J. (2012). *FCC chairman: Digital textbooks to all students in five years*. Retrieved from http://www.fcc.gov/blog/fcc-chairman-digital-textbooks-all-students-five-years

Yerushalmy, M. (2016). Inquiry curriculum and e-textbooks: Technological changes that challenge the representation of mathematics pedagogy. In M. Bates & Z. Usiskin (Eds.), *Digital curricula in school mathematics* (pp. 87–106). Charlotte, NC: Information Age.

Zhao, Y., Zhang, G., & Lai, C. (2010). Curriculum, digital resources and delivery. In P. Peterson, E. Baker & B. McGaw (Eds.), *International Encyclopedia of Education* (pp. 390–396). Oxford, United Kingdom: Elsevier Ltd.

CHAPTER 12

A DESIGN EXPERIMENT OF A DEEPLY DIGITAL INSTRUCTIONAL UNIT AND ITS IMPACT IN HIGH SCHOOL CLASSROOMS[1]

Alden J. Edson

MOVING TOWARDS "DEEPLY DIGITAL" CURRICULA IN SCHOOL MATHEMATICS

In a technological society with a globalization of Internet availability and the widespread use of portable devices, it is possible to foresee new ways of learning involving present and emerging digital technologies. In turn, these new ways of learning suggest new designs for mathematics curriculum. "Just as information technologies are reshaping almost every phase of contemporary scientific, technical, business, and personal life, they are challenging long-standing patterns of learning and teaching and the curricula of K–12 schools" (Fey, 2009, p. 1). A recent survey shows that "over two-thirds of the teachers reported using digital materials as teacher resources either once or twice a week or almost every day" (Davis, Choppin, Roth McDuffie, & Drake, 2013, p. 10) in an effort to align district-adopted curriculum programs to the Common Core State Standards (National Governors Association Center for Best Practices & Council of Chief State School Officers, 2010). Although many of these electronic resources include lessons and activities found on websites, many publishers and education software companies are releasing comprehensive programs designed to supplement or supplant print textbooks in the form of digital curricula (Choppin, Carson, Borys, Cerosaletti, & Gillis, 2014).

There is a dearth of research reporting on the digital instructional materials in school mathematics. The design of digital versions of curriculum programs typically consists of simple web pages or static electronic conversions of print materials. Choppin et al. (2014) report on a typology developed for analyzing design features relating to students' interactions with the program, curriculum use and adaptation, and analysis of assessment systems. Results show there are two main types of digital curriculum programs in mathematics: individualized learning programs and digitized versions of conventional textbooks. The results from Chopin et al. (2014) indicate that, although digital materials are still in their infancy, design features identified in the field of educational technology as potentially transformative for teacher classroom practices and student learning of mathematics do not leverage present and emerging technologies.

"Deeply digital" instructional materials hold promise as the next stage in the continuum of design and development of digital instructional resources in school mathematics. As Roschelle (2011) describes in a podcast, "A digital textbook is not just taking an image of a textbook and putting it on an electronic device to read it. A digital textbook is something that takes school content and takes advantage of the properties of digital media to produce that school content in a completely new way that's much stronger for learning." Deeply digital instructional materials refer to instructional materials with design features that leverage the affordances of digital technologies for teaching and learning mathematics, including (but not limited to) embedded models, simulations, and visualizations of mathematical topics; coherent and integrated resources through an embedded network of hyperlinks; and feedback support through the interconnectivity between student and teacher materials (Dorsey, 2011). Examples of the next generation digital instructional materials include products featured at the Center for the Study of Mathematics Curriculum Third International Curriculum Conference: Mathematics Curriculum Development, Delivery, and Enactment in a Digital World. Although these emerging digital instructional materials were not included in Choppin et al. (2014), they are reported on in this volume.

A PROTOTYPE OF A DEEPLY DIGITAL UNIT ON BINOMIAL DISTRIBUTIONS AND STATISTICAL INFERENCE

In 2010, the National Science Foundation funded a three-year research and development project[2] to design, create, field test, refine, and bring to publication Transition to College Mathematics and Statistics (TCMS; Hirsch et al., 2015), a fourth-year high school mathematics course intended for students who do not plan in college to major in engineering or the mathematical, physical, or biological sciences. Key features of TCMS include: (a) a problem-based, inquiry-oriented, technology-rich approach to mathematics and statistics; (b) student-centered investigations promoting active learning, teamwork, and communication; (c) real-world applications and mathematical modeling contexts for developing and connecting important mathematical content and mathematical practices; (d) curriculum-embedded software, TCMS-Tools, to support student inquiry, mathematical modeling, and problem solving; and (e) a flexible design to enable TCMS to be used following a conventional single-subject curriculum or with an integrated curriculum program. Both a print and highly interactive digital version of the student text have been completed, as well as an emerging deeply digital instructional unit to serve as a proof of concept of what could be possible in high school mathematics. The design features of the deeply digital materials were selected based on the premise of providing a vision of what curriculum materials could look like in a digital world.

The research reported in this chapter extends the work of a pilot design experiment involving the Binomial Distributions and Statistical Inference unit by creating and testing a more deeply digital unit. The purpose of the research was to (re)design, iteratively develop, test, and evaluate a deeply digital instructional unit focusing on binomial distributions and statistical inference. In this chapter, the research reports on how, if at all, the design features of the deeply digital unit transformed the roles of both students and teacher in a deeply digital learning environment. The overall methodological scheme of the larger research project is a design experiment (Brown, 1992; Collins, 1992, 1999; Design-Based Research Collective, 2003) that is similar in many ways to other design research approaches, such as developmental research (Richey, Klein & Nelson, 2004), design research (Edelson, 2002), and formative research (Reigeluth & Frick, 1999).

The unit objectives for the "Binomial Distributions and Statistical Inference" unit were to: (a) develop the binomial probability formula and know when it can be used; (b) compute and interpret a p-value, deciding whether the number of successes in a binomial situation is statistically significant; (c) understand and apply basic ideas related to the design and interpretation of surveys, such as random sampling and bias, critically analyzing surveys and polls in everyday life and as reported in the media; and

(d) compute and interpret the margin of error and the confidence interval for a population proportion. The unit is organized around four focused yet connected lessons, each consisting of one to three clusters of specific multi-day student investigations. The student focusing questions for the investigations that compose the unit are shown in Figure 12.1.

The process of the (re)design and iterative development, shown in Figure 12.2, began with the development of the print unit. Iteration I consisted

---
Lesson 1: *Binomial Distributions*

Investigation 1: Rules of Probability and Binomial Situations
- *How can rules of probability help you analyze a binomial situation?*

Investigation 2: The Binomial Probability Formula
- *How can you find the probability of getting a specified number of successes x in a binomial situation with n trials and probability p of success on each trial?*

Investigation 3: Statistical Significance
- *How can you use technology to find out whether the number of successes in a binomial situation is statistically significant?*

Lesson 2: *Sample Surveys*

Investigation 1: Trustworthy Surveys
- *What are the characteristics of a trustworthy survey?*

Investigation 2: How Surveys Can Go Wrong
- *What are the characteristics of an untrustworthy survey?*

Lesson 3: *Margin of Error From Sample to Population*

Investigation 1: Variability from Sample to Sample
- *How far is the proportion of successes in a random sample likely to be from the proportion of successes in the population from which it was taken?*

Investigation 2: The Margin of Error
- *How can you estimate the sampling error when you do not know the proportion p of successes in the population?*

Investigation 3: Interpreting a Confidence Interval

  How is a 95% confidence interval computed and how should it be interpreted?

Lesson 4: *Looking Back*

Figure 12.1 Overview of the "Binomial Distributions and Statistical Inference" unit.

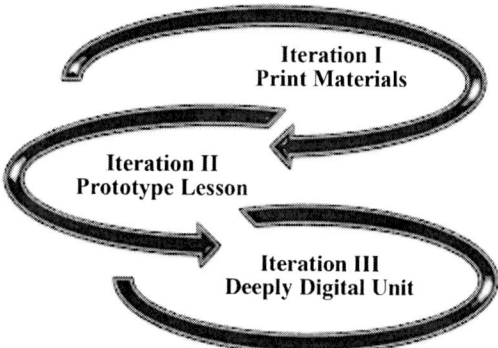

Figure 12.2 Overall (re)design and iterative development process.

of six key steps: (a) identifying the key ideas of binomial distributions with application to statistical inference and then systematically outlining the scope, sequence, and development of those ideas with advice from mathematicians, statisticians, and experienced teachers; (b) researching engaging and realistic contexts for investigations and problems developing the intended statistics; (c) developing draft student and teacher materials with local piloting of early iterations; (d) designing and developing formative and summative assessments of student conceptual understanding and proficiency with the targeted content; (e) field testing and evaluation of draft materials; and (f) revisions and refinements of the tested materials prior to publication.

In Iteration II, the pilot study, the print unit was adapted for one digital version of the first lesson (consisting of three multi-day student investigations). The purpose of the pilot design experiment was to investigate the different ways students used the pedagogical and tool features as they solved problems and constructed their learning. Modifications made to the prototype lesson were based on feedback from classroom observations and digital artifacts, student interviews, and student and teacher surveys.

The research reported in this chapter is part of a larger research project (Edson, 2014). It is shown as Iteration III in the process and was scaled up to four lessons in the TCMS "Binomial Distributions and Statistical Inference" unit. Figure 12.3 shows the unit opener and the digital shell for a lesson and investigation of the digital materials. The pedagogical and tool features of the deeply digital instructional unit include: (a) embedded audio and video clips, (b) curriculum-embedded mathematical and statistical software, (c) accessibility features such as text-to-speech and translation software, (d) embedded digital collaborative notebooks, and (e) learner-controlled scaffolding linked to open-ended problems.

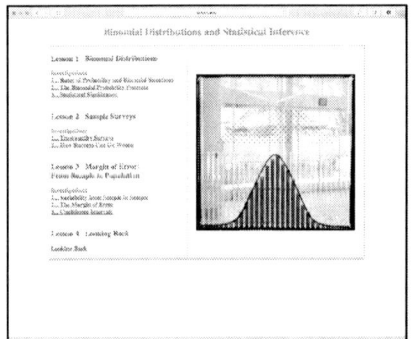 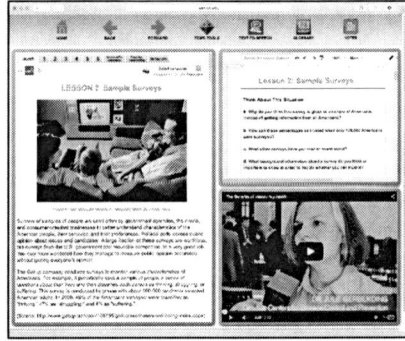

Figure 12.3 Unit opener and digital shell of the digital materials.

The four main processes of the larger research project that describe the design experiment include: (a) an explication of the learning progression intended by the print unit; (b) a (re)design and iterative development of a deeply digital instructional unit organized by lessons consisting of clusters of specific, yet connected, student investigations; (c) an enactment of the developed set of four lessons in a student-centered digital learning environment; and (d) a retrospective analysis on the design, development, and enactment of the deeply digital instructional unit. The research was conducted within a fourth-year high school mathematics classroom that contained two interactive whiteboards connected to a single audio system, remotely controlled by the teacher's laptop and tablet. Twenty participating students were placed into groups by the teacher based upon student attendance, mathematical ability (high, medium, low achieving) as measured in part by pre-assessment findings, technology use and needs, and willingness to participate in the study. For the duration of the study, students were in heterogeneous groups consisting of four students, with at least one student identified as at a different ability level. Every student in the school had access to the digital materials on their laptop at school and at home for the entire study. Every student had wirelessly connected to the school server for Internet access.

Data sources included classroom observations and digital artifacts, student and teacher interviews, and student assessments and surveys. Classroom digital artifacts consisted of student work from their digital collaborative notebooks, photos of relevant board work, and scans of paper-and-pencil work including lesson quizzes. Observation field notes included information about the classroom setting, role of the teacher, role of the student, and use of the materials; these were analyzed for main themes. Pre- and post-unit assessments were administered. Semi-structured student and teacher interviews were conducted near the end of the study. These focused on the mathematical tasks, the role of the teacher, the role of the students, and their interactions with the digital unit. Student surveys captured feedback on use of the digital features of the prototype unit and were administered electronically after each lesson (with the exception of the fourth lesson, a review and synthesis lesson).

USE OF THE FEATURES OF THE DEEPLY DIGITAL MATERIALS AND ROLES OF STUDENTS AND THE TEACHER

This section reports on the digital design features of the deeply digital instructional unit and how, if it all, the features transform the roles of both students and teacher in a deeply digital learning environment. Descriptions of the design features of the digital materials and related student use of each of the five different digital design features organize this section. Following this description is a discussion of the identified changes in the role of the teacher.

Student Use of the Embedded Audio and Video Clips

One design feature of the digital materials includes the embedded audio and video clips. These clips are accessed by a hyperlink located within the text of the problem situation (typically under a photo related to the problem context) and are viewed in the lower right-hand frame of the shell of the digital materials. The video clips are short in duration (no more than five minutes), contain both audio and video, incorporate relevant visual material, and are embedded throughout the problem situations. For instance, Figure 12.4 shows the launch of a problem situation and a set of related questions for students to discuss in a whole-class setting. Students can access a related audio clip about the *Avery v. Georgia* court case by clicking on the hyperlink located under the photo.

Results indicated that students accessed the embedded audio and video clips and related them to the different problem situations. Students indicated that they accessed them because they were curious to learn more about the different problem situations. For instance, one student stated, "If you're just starting a lesson, the video and audio clips can help you to know what the lesson will be about and can help you to figure out problems within the lesson." Others spoke to student engagement and motivation. For example, "If there is a scenario or problem relating to a topic and we are unfamiliar with it, we could watch the video to learn more about it and expand our knowledge as a whole. Also, I tend to get curious and click on the video anyway." The teacher indicated that the embedded audio and

> The ethnicity, gender, age, and other demographic characteristics of juries have been of great interest in some trials in the United States. When the composition of a jury does not reflect the demographic characteristics of the surrounding community, doubts about fairness of the jury selection process and legal challenges can arise.
>
> Although juries are not selected solely by chance, comparing the actual jury to the composition of juries that would occur if jurors were selected at random can tell lawyers whether there are grounds to investigate the fairness of the jury selection process.
>
> Listen and read about Avery v. Georgia [345 U.S. 559 (1953)] via ACLU Montana
>
> An historic case concerning jury selection, Avery v. Georgia, was brought to the U.S. Supreme Court in 1953. A jury in Fulton County, Georgia had convicted Avery, an African-American, of a serious felony. There were no African-Americans on the jury. At the time, there were 165,814 African-Americans in the Fulton County population of 691,797. The list of 21,624 potential jurors had 1,115 African-Americans. A jury pool of 60 people was selected, supposedly at random, from the list of potential jurors. (However, the names of black and white jurors had been written on different colored slips of paper.) This jury pool, from which the 12 actual jurors were selected, contained no African-Americans.

Think About This Situation

Think about the demographics of Fulton County, Georgia and of the jury selected for the trial of James Avery.

- (a) If 12 jurors were selected at random from the people in Fulton County, how can you compute or estimate the probability that there would be no African-Americans on the jury? Generate as many methods as you can.
- (b) Do you think that having only 1,115 African-Americans on the list of potential jurors reasonably can be attributed to chance alone or should the lawyers look for another explanation? What strategies could you use to support your choice?
- (c) The jury pool of 60 people was selected from the list of 21,624 potential jurors. Can getting no African-Americans in a jury pool selected from these potential jurors reasonably be attributed to chance alone or should the lawyers look for another explanation? What strategies could you use to support your choice?
- (d) The U.S. Supreme Court overturned Avery's conviction. Describe the statistical evidence that you think might have been used by Avery's lawyers.

Figure 12.4 Launch for a problem situation and set of related questions. (Adapted from Hirsch et al., 2015, and photos used with permission of McGraw-Hill, 2015).

video clips provided support for an alternate format to launching various problem situations. The teacher contrasted the experience between using the digital materials and using the conventional print materials. In environments using the print materials, either the students or the teacher read aloud the text describing the problem situation before discussing the set of related questions. The teacher indicated that the short multimedia clips help to motivate and engage students in the different problem situations in the different medium.

Student Use of the Curriculum-Embedded Mathematical and Statistical Software

Another design feature of the digital materials is curriculum-embedded mathematical and statistical software. Affordances of the mathematical and statistical software in the deeply digital environment included the ease to embed screenshots from laptop computers and use of TCMS software tools to solve problems. Students can use TCMS-Tools (Keller, 2014), a Java-based suite of both general-purpose (e.g., spreadsheet, a computer algebra system, interactive geometry, data analysis, and simulation tools) and custom tools designed for specific mathematical and statistical topics. For example, the TCMS-Tools custom app, Binomial Distributions, allows users to vary the number of successes and the probability of a success on each trial to see how the shape, center, and spread of binomial distributions are affected. Students are also able to conduct trials and compare theoretical models of distributions to simulation results, as shown in the different snapshots of Figure 12.5.

Results indicated that students embedded TCMS-Tools screenshots into their digital collaborative notebooks as they used spreadsheets, graphs, and computer algebra systems (CAS) to solve binomial distribution problems. This occurred primarily when students were exploring problems in groups and wanted to digitally record the commands entered into the software, to publically share the output of their work, or to access their work at a later time. Students generally said that the software provided easier and quicker access to a suite of tools, and the software was easier to use on their laptop computers because they have larger screens than their graphing calculator.

Student Use of the Accessibility Features

The digital materials provide amplification around accessibility features such as options to adjust sound, change the size of the text, control graphics, and gain access to outside tools embedded in the materials, such as

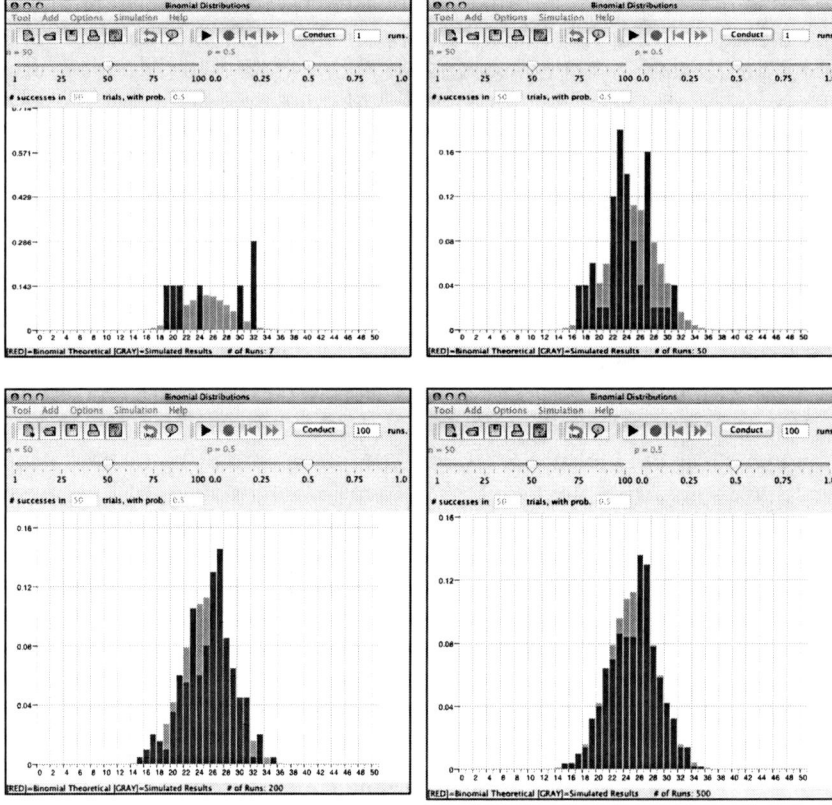

Figure 12.5 Snapshots of a TCMS-Tools custom app focusing on binomial distributions.

text-to-speech software that reads aloud the written text and translation software that translates from one written language to another. Each investigation problem is contained on its own hyperlinked tab. Figure 12.6 shows snapshots of the scroll-over definition feature and glossary definitions for the term "independent trials."

The research results indicated that students used the glossary of key terms and scroll-over definitions to recall definitions and examples. A total of 31 vocabulary terms were included in the glossary for the instructional unit. Each time the vocabulary term appeared as text in the digital materials, students could access the scroll-over definition for that term by placing their cursor over its location. Of the twenty students, 85% indicated that the glossary and scroll-over definition features were *somewhat* or *very helpful*. Also, 95% indicated that the glossary and scroll-over definitions would be useful for students taking the TCMS course at other schools. When

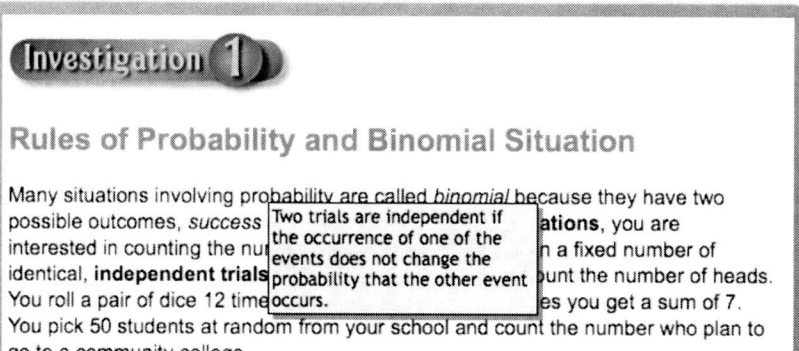

Figure 12.6 Snapshots of scroll-over definition and glossary definition for independent trials.

providing suggestions for how the digital materials could be improved, students generally said they wanted more terms considered for the scroll-over definition and glossary features. Students solved problems together in their groups using the "problem on its own tab" feature and accessed embedded information using the hyperlinks. Students indicated that the embedded hyperlinks, glossary of key terms, scroll-over definitions, text-to-speech, print feature, adjustment of text size, and "problem on its own tab" are valuable features for accessing resources for investigation problems and homework tasks.

Student Use of the Embedded Digital Collaborative Notebooks

Another design feature of the digital materials is the use of embedded digital collaborative notebooks. The digital collaborative notebook feature makes use of Google Drive. Students can communicate and collaborate in a synchronous and shared learning environment designed for students individually, in small groups, or as a whole-class to organize their work. As an illustration, Figure 12.7 shows student work from four different students' screens on the same investigation problem, whereas Figure 12.8 shows student work from a group of three students, where each color represents a

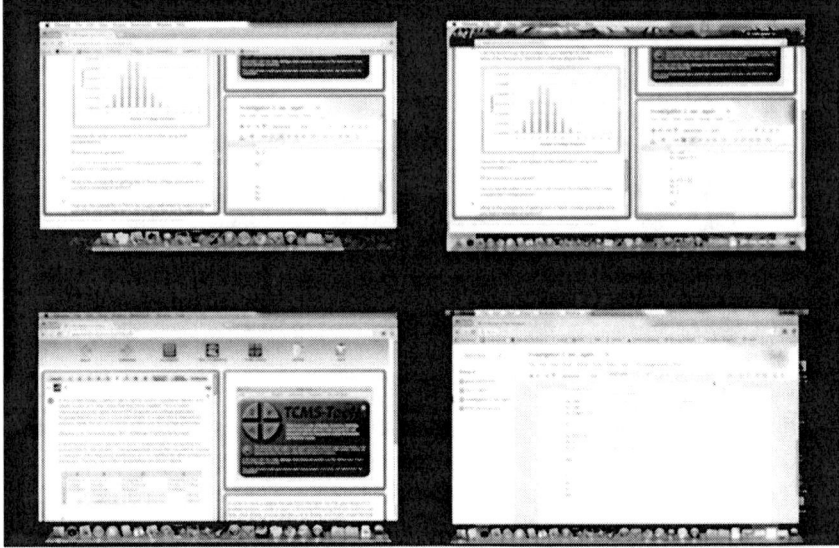

Figure 12.7 Digital collaborative notebook feature and related student work of four students.

different student. With the digital collaborative notebook feature, student work is automatically archived digitally and is available outside of class.

Research results indicated that students contributed to, revised, and reflected on their own thinking and the thinking of their group mates. They incorporated tables, sketches, comments, and photos into their digital collaborative notebooks when solving problems. Students stayed together with their group mates before moving to additional investigation problems while using their digital collaborative notebooks. Students accessed their completed group work on their digital collaborative notebooks outside of class to help them solve individual homework problems.

Student Use of the Open-Ended Problems Linked to Learner-Controlled Scaffolding

Another design feature of the digital materials is that investigation problems and homework tasks are organized around open-ended problems linked to learner-controlled scaffolding, as illustrated in Figure 12.9. Students can click on the blue hyperlinks, "Not sure how to get started?" and "Be persistent. If necessary, consider thinking about this." that reveal alternate or sub-problems connected to the initial, more open-ended problem.

Figure 12.8 Example of work from three students.

Research results indicated that students solved complex and more open-ended problems as compared to when they used the print materials. Groups of students used the learner-controlled scaffolding in different ways when they had difficulties on an investigation problem or homework task. In terms of how the scaffolding was used, groupmates would collectively use the learner-controlled scaffolding, discuss their thinking about the problem, and then write their thinking in the digital collaborative notebooks. For other groups, one or two groupmates would use the learner-controlled scaffolding individually and incorporate the idea of the prompt into their group discussion about the investigation problem. One student said, "The prompts help those who are stuck on a particular problem, by having these prompts available they help those to get on the right path, without giving away the answer." In addition, students used the learner-controlled scaffolding as a prompt for reflecting on their thinking. For instance, "If you have done the problem and [are] not sure if you have done it right, then click

Number on Green Die

		1	2	3	4	5	6
Number on Red Die	1	1,1	1,2	1,3	1,4	1,5	1,6
	2	2,1	2,2	2,3	2,4	2,5	2,6
	3	3,1	3,2	3,3	3,4	3,5	3,6
	4	4,1	4,2	4,3	4,4	4,5	4,6
	5	5,1	5,2	5,3	5,4	5,5	5,6
	6	6,1	6,2	6,3	6,4	6,5	6,6

a. Using the information displayed in the sample space above, create a representation that shows the probability distribution for the sum of the two dice.

 🔵 *Not sure how to get started?*
What are the horizontal and vertical axes in a graphical representation? What are the column headers in a table? Is it better to use simplified fractions, unsimplified fractions, or decimals?

b. Write a description of the shape of the probability distribution. Formulate a probability question that can be answered using your representation of the distribution. Discuss your questions and answers in your group.

Figure 12.9 Illustration of an open-ended problem linked to learner-controlled scaffolding. (Adapted from Hirsch et al., 2015, and photos used with permission of McGraw-Hill, 2015).

on the [scaffolding], that [the blue button links], and make sure that it still makes sense using the next context of the prompt."

The Role of the Teacher in a Deeply Digital, Student-Centered Learning Environment

There was a shift in the role of the teacher in a deeply digital, student-centered learning environment that further supported the teacher in guiding, questioning, and facilitating the students to do and make sense of mathematics. When facilitating whole-class discussion for lesson and investigation launches, problems and related student questions were amplified through use of embedded audio and video clips that targeted the mathematical/statistical and cultural suppositions inherent in contextual problem situations. During student group explorations of investigation problems, rather than completing a static sequence of problems and their scaffolding, students took ownership of their own learning by using the learner-controlled scaffolding without teacher intervention. Student use of the learner-controlled scaffolding supports teachers in fostering an environment that optimizes the level of challenge for students while simultaneously providing students with a mechanism for accessing entry into investigation problems and homework tasks. If and when students request help from the teacher, the nature of the student questions focus more on understanding the underlying content of the problem situation than on getting started with the initial open-ended problem. During whole-class discussions that follow investigation work, where students have opportunities to summarize and explain their reasoning about mathematical and statistical ideas, the use of the digital collaborative notebook provides teachers and students with a quick and flexible way to share student work on investigation problems without the additional steps and time needed in classrooms using print materials.

CONCLUSION

The purpose of this chapter was to describe how, if at all, the design features of the deeply digital unit transformed the roles of both students and teacher in a deeply digital learning environment. The digital design features of a deeply digital instructional unit consisted of five pedagogical and tool features: (a) embedded audio and video clips, (b) curriculum-embedded mathematical and statistical software, (c) accessibility features such as text-to-speech and translation software, (d) embedded digital collaborative notebooks, and (e) learner-controlled scaffolding linked to open-ended problems. The design features were articulated in terms of how students and teachers used these features and their perspectives for how they supported teaching and learning. Findings suggest that if students and teachers use deeply digital materials, such as those reported in this study, classroom instruction may potentially be transformed. Findings also support a conclusion that deeply digital, problem-based, inquiry-oriented, instructional mathematics materials may provide more support for teaching and learning than typical digitized versions of conventional textbooks currently in use, as described by Choppin et al. (2014).

NOTES

1. The research and its interpretation reported in this chapter are based on the author's doctoral dissertation completed at Western Michigan University under the direction of Christian R. Hirsch and Steven W. Ziebarth.

2. This research was supported in part by the Transition to College Mathematics and Statistics project with funding from the National Science Foundation Grant DRL-1020312. All opinions and analysis expressed herein are those of the author and do not necessarily represent the position or policies of the Foundation. The author gratefully acknowledges the work of Ann Watkins, California State University–Northridge, in developing the print version of the Binomial Distributions and Statistical Inference unit and that of James Laser, Western Michigan University, in coding the initial shell for the digital unit.

REFERENCES

Brown, A. L. (1992). Design experiments: Theoretical and methodological challenges in creating complex interventions in classroom settings. *Journal of the Learning Sciences, 2*(2), 141–178.

Choppin, J., Carson, C., Borys, Z., Cerosaletti, C., & Gillis, R. (2014). A typology for analyzing digital curricula in mathematics education. *International Journal of Education in Mathematics, Science, and Technology, 2*(1), 11–25.

Collins, A. (1992). Toward a design science of education. In E. Scanlon & T. O'Shea (Eds.), *New directions in educational technology* (pp. 15–22). New York, NY: Springer.

Collins, A. (1999). The changing infrastructure of education research. In E. C. Lagemann & L. S. Shulman (Eds.), *Issues in education research: Problems and possibilities* (pp. 289–298). San Francisco, CA: Jossey-Bass.

Davis, J., Choppin, J., Roth McDuffie, A., & Drake, C. (2013). *Common Core State Standards for Mathematics: Middle school teachers' perceptions*. Retrieved from http://www.warner.rochester.edu/files/warnercenter/docs/commoncoremathreport.pdf

Design-Based Research Collective. (2003). Design-based research: An emerging paradigm for educational inquiry. *Educational Researcher, 32*(1), 5–8.

Dorsey, C. (2011). *Perspective: Defining a deeply digital education.* Retrieved from http://concord.org/publications/newsletter/2011-fall/perspective

Edelson, D. C. (2002). Design research: What we learn when we engage in design. *Journal of the Learning Sciences, 11*(1), 105–121.

Edson, A. J. (2014). *A deeply digital instructional unit on binomial distributions and statistical inference: A design experiment* (Unpublished doctoral dissertation). Western Michigan University, Kalamazoo, MI.

Fey, J. (2009). *Considering the future of K-12 STEM curricula and instructional materials: Stimulating and supporting new developments.* The Center for the Study of Mathematics Curriculum Workshop Series on STEM Curriculum & Instructional Design. Retrieved from http://mathcurriculumcenter.org/conferences/stem/index.php

Hirsch, C. R., Hart, E. W., Watkins, A. E., Ritsema, B. E., Fey, J. T., Keller, B. A., . . . Laser, J. K. (2015). *Transition to college mathematics and statistics*. Columbus, OH: McGraw-Hill School Solutions.

Keller, B. (2014). *TCMS-Tools* [Software]. Kalamazoo, MI: The Transition to College Mathematics and Statistics Project, Western Michigan University.

National Governors Association Center for Best Practices & Council of Chief State School Officers. (2010). *Common Core State Standards for Mathematics.* Washington, DC: Authors.

Reigeluth, C. M., & Frick, T. W. (1999). Formative research: A methodology for creating and improving design theories. In C. M. Reigeluth (Ed.), *Instructional-design theories and models: A new paradigm of instructional theory* (Vol. 2, pp. 633–651). Mahwah, NJ: Lawrence Erlbaum.

Richey, R. C., Klein, J., & Nelson, W. (2004). Developmental research: Studies of instructional design and development. In D. Jonassen (Ed.), *Handbook of research for educational communications and technology* (2nd ed.; pp. 1099–1130). Mahwah, NJ: Lawrence Erlbaum.

Roschelle, J. (Producer). (2011). *Digital textbooks: New visions for interactive learning* [Podcast]. Retrieved from http://www.sri.com/news/podcasts/JeremyRoschellePodcast.html

CHAPTER 13

KEEPING AN EYE ON THE TEACHER IN THE DIGITAL CURRICULUM RACE

Janine Remillard

Digital curriculum resources provide a range of possibilities for organizing content and engaging learners in exploring mathematics in new and innovative ways. They also present new challenges to teachers. Many of the presentations and discussions at the conference focused on digital curriculum designs for students. I consider the possibilities and demands of digital curriculum resources from the teachers' perspective.

The focus of my research is teachers as users and re-sourcers[1] of curriculum materials and tools. For over 20 years, I have studied teachers' interactions with curriculum materials, focusing, primarily, on one question: Can curriculum materials serve as a positive tool for change in teaching?

This question is a personal one for me; my teaching was fundamentally changed by a curriculum program. As a new teacher in the 1980s, I came across an innovative mathematics program called Comprehensive School Mathematics Program (CSMP) (Herbert, 1984). It was unlike any textbook I had ever seen. Using it helped me engage my students in reasoning about important mathematical concepts and ideas. It suggested questions to pose to students that required them to think about underlying meanings and not simply find answers. Following the suggestions in the teacher's guide helped me establish conversations in my math class.

CSMP also pushed me to think about mathematical ideas in new ways. The visual representations used to illustrate concepts and relationships deepened my understanding. And, the way three strands (number, function, and operation) formed the spiral structure of the curriculum gave me new insight into how they fit together. Through using the curriculum, I began to think about numbers, functions, and operations more conceptually and as interconnected.

Another way CSMP was different from other textbooks of the time was in how the lessons were structured. Rather than revolving around a set of textbook pages, the lessons outlined in the teacher's guide were very interactive and involved work on problems. To plan, I had to spend time each night reading the text, working through the problems myself, and imagining what my students would do. Reading, planning with, and modifying—re-sourcing—the curriculum program provided me with a new way to organize the mathematics I was teaching and significantly changed my teaching.

To my surprise, once we adopted the new program, a number of colleagues at my school were much less enthusiastic. They did not all welcome the unfamiliar approach, had difficulty navigating some of the representations, and made some substantial revisions, such as untangling the spiral structure, teaching each strand fully before moving onto the next.

This experience informed and influenced my research. In my research on teacher development and change, I have seen that curriculum resources can support teachers when trying new things and, at the same time, that teachers call the shots when it comes to how they use instructional resources. My aim has been to understand teachers' interactions with curriculum materials. Ultimately, I want to consider how these resources and other tools can be designed to support teachers to design and enact curriculum. What features do the resources need to have? What are the implications for how teachers use them? I have studied these questions, primarily, through print materials.

In this chapter, I consider implications of this research for teachers' use of digital resources. I start by clarifying some basic assumptions about teachers and curriculum resources that guide my work. Then, I describe some key characteristics of digital curriculum resources in an online environment, considering the demands teachers navigate. I also propose specific areas where research is needed, based on what we know about teachers' use of curriculum resources and the digital curriculum environment. I want to urge digital curriculum developers and those who work with teachers to not lose sight of teachers and teaching in this fast-moving landscape.

ASSUMPTIONS

My comments and speculations are based on several related assumptions about curriculum resources and educational practice in general. First, teachers matter. Clearly teachers' roles will need to change with the evolution of digital tools, particularly when these tools are designed to engage students directly and often individually. Still, teachers are critical guardians of the learning process. The conference was held at the University of Chicago, so it is fitting to recall Dewey's (1897) assertions about the role of the teacher in shaping interactions between the child and the curriculum. The teacher is to "select the influences which shall affect the child" and "assist him in properly responding to these influences" (within pp. 77–80). In my view, the most critical role teachers play, particularly in a digital environment, is ensuring equity of opportunity and access to the learning resources afforded by digital tools.

Second, resource use is best understood from a socio-cultural perspective. Vygotsky's (1978) analysis of the mediating role tools play in human activity provides a framework for examining teacher–resource interactions. The practices we engage in are inseparable from the tools we employ (and produce through the process) and are deeply rooted in the particular context in which they are used.[2] From this perspective, we know that, along with teachers, tools matter. As I reflect on my experience as a young teacher, detailed previously, I am reminded of a quote attributed to Richard Buckminster Fuller, shared by David Moursund (2016) at the conference: "If you want to teach people a new way of thinking, don't bother trying to teach them. Instead give them a tool, the use of which will lead to new ways of thinking." A new curriculum tool certainly changed my way of thinking about math and how to teach it.

Third, for any type of curriculum resource and for any teacher, using a resource is never a matter of simply taking it up and following it. When teachers "use" curriculum resources, they are interpreting, authoring, appropriating, adapting, framing, and reframing the content.[3] This perspective reflects the notion of documentational genesis, a term offered by Gueudet and Trouche (2009) to describe teachers' work with documents and the production of new documents that result from these activities. This theoretical perspective holds that documents are situated within schemes of usage assumed by the teacher. As such, using resources necessarily involves re-sourcing them for one's specific purpose and context.

KEY CHARACTERISTICS: TWO TYPES OF DIGITAL ENVIRONMENTS

As mentioned at the beginning, my aim is to consider the possibilities and demands of digital curriculum resources from the teacher's perspective.

What types of teaching and learning do these resources make possible? What is expected of the teacher? What demands do these resources place on them? In order to do so, we need to take a closer look at the characteristics of these resources.

In recent years, I have observed that teachers find themselves navigating two types of digital environments. Both need to be understood. The first type is shaped by digitally-designed curriculum resources, such as the tools, programs, and interfaces that were the focus of the presentations and conversations during the conference. These resources are deliberately designed to serve as instructional tools for teachers or students. The second type of environment is shaped by the vast collection of potential curriculum resources available to teachers, often marketed to them, through the Internet. In many parts of the world, instructional material production and dissemination is a commercial enterprise. Digitization and the web have expanded the availability and access of new resources at all levels of the system. These evolutions have influenced the quality and quantity of instructional materials available and created new challenges for teachers to navigate.

Characteristics of Designed Digital Resources

Designed digital resources have both structural and use characteristics. One characteristic, made possible by digital technologies, is that they can assume a nonlinear structure. Rather than offering a single path through a particular terrain, they often take the form of constellations with multiple possible connections and routes. Digital resources are also opaque (Usiskin, 2013). The contents and structure of print materials are typically much more transparent and visible than digital. In the same way, digital resources have the possibility to be layered, allowing the user the option of delving deeper into one item through a series of embedded links. Finally, digital resources are connection rich. Some resources explicitly make connections across curricular components and with the outside world. Although print resources also allow for similar connections, the ability to provide links to internal and external resources makes this characteristic more pronounced in digital resources.

In terms of use characteristics, digital resources are flexible and often designed to be adapted by the user. In this way, they can be personalized for and by individual students or teachers. Further, they allow for multiple modes of engagement by the user. Rather than simply reading and interacting with print material, users of digital resources are more likely to read, watch video, manipulate, create, and respond; in this way, digital resources are more likely than print materials to involve interactivity.

As others have pointed out, these characteristics make it possible for the teacher to design and make available rich learning opportunities for students.[4] With these opportunities come increased demands for re-sourcing in a highly complex and opaque environment. I return to this challenge in the final section.

Characteristics of Resources in a Digital World

Determining the characteristics of resources in today's digital world is not a straightforward matter. I have in the past tried to identify characteristics of this vast landscape of resources, but it is difficult. Variation is the norm. The resources vary in grain size and comprehensiveness; I avoid using the term "quality" because it suggests we have clear criteria for determining quality of digital resources and we do not. It is fair to say, however, that these resources vary in the curricular aim and nature of learner experience they offer. They even vary in the extent to which they make use of digital capacities. Some resources are more "deeply digital," to use a term offered by Chad Dorsey;[5] others are static documents published in an electronic format. Most fall somewhere between these two ends.

I have observed two patterns across these resources. First, they tend to be packaged in discrete units. Even when designed as part of a comprehensive system, it is the activity or task, the applet, the worksheet, the practice test that is often appropriated and re-sourced by teachers. Second, they are widely available to and directly accessed by teachers and used for various purposes.[6] I have thought of this arrangement as akin to the wild west or the wild, wild west, representing an alternate meaning of the "www" in URLs.

My aim is not to critique this state of affairs, but to simply identify it as a reality of teachers' curriculum design work. I also argue that design and research might attend more to how this landscape of resources might be navigated productively.

RESEARCH ON TEACHERS' USE OF DIGITAL CURRICULUM RESOURCES: WHAT IS NEEDED?

Given the characteristics of digital resources and digital environments previously discussed, I now turn to critical areas where research is needed. In this section, I propose three questions that researchers might explore. Each question has been (and is presently being) explored with teachers using print-based curriculum resources. I consider issues raised by the unique characteristics of digital resources.

What Is Involved in Using Curriculum Resources Effectively?

For some, the question of effective use of curriculum resources has focused on teachers using them as intended. Others focus on use that produces the most desirable learning by students. Given the flexible and customizable nature of many digital curriculum resources, it is fair to say that "intended use" is either impossible to specify or an unhelpful measure. With this point in mind, the critical question becomes: What is involved in using digital resources in ways that promote desired student learning? Research that examines teachers' interactions with and use of digital resources is needed to understand the work of teachers in this terrain.

Researchers, focusing primarily on print resources, have argued for understanding teachers' use of curriculum resources as a complex and sophisticated activity that involves making meaning of the resource (or multiple resources) and adapting or reformulating it (either conceptually or structurally) for use in one's own classroom. Some of the ways curriculum use has been conceptualized include participation (Remillard, 2005), documentational genesis (Gueudet & Trouche, 2009), and resourcing (Pepin, Gueudet, & Trouche, 2013). When incorporating digital environments, some have used the term "curate" to describe teachers' work. This term captures the need to navigate a vast array of digital resources available to teachers and students and the fact that many resources are designed for individual engagement by students. Like the curator of a museum collection, teachers collect, organize, and design learners' paths through a compilation of tools. Usiskin (2010) used the metaphor of a tour bus driver to describe the teacher's role in this area. If curriculum designers are engineers building roads through the mathematical domains and curriculum supervisors are like travel agents planning the trip, "teachers are the bus drivers on these roads" (pp. 25–26). They select the speed of travel and decide which sites to spend time at and which to skip.

A metaphor I find particularly relevant to the work of navigating curriculum resources in digital environments is that of cartographer. Cartography is the study and practice of map *making*. This metaphor borrows from Usiskin's (2010) travel image, but acknowledges that much of today's rapidly expanding landscape of digital resources is not yet fully known. A mass of roads has been built, both superhighways and tiny lanes, but the choices of how to travel them are vast and few trips have been preplanned. Moreover, in many school districts in the United States, the position of curriculum supervisor, the likely trip planner, has been cut from the budget. I find cartography an apt metaphor for curriculum use in digital environments because mapping involves both representing a domain and setting one or more paths through it in a way that allows others to understand that space more fully. According to several sources, mapmakers must address the following challenges:

1. Determine which traits of a domain will be mapped.
2. Represent the region using mapping tools available.
3. Generalize by eliminating features of the region that are not relevant to the purpose of the map and determine an appropriate level of complexity for those features that will be included.
4. Design the elements of the map using standard conventions (or proposing novel ones) to convey the meaning to the intended audience.

Considering the complexity of these tasks leads us to a second research question: What types of knowledge and abilities are required of teachers to use curriculum resources effectively?

What Types of Knowledge and Abilities Are Involved in Using Curriculum Resources Effectively?

Brown (2009) offered a framework for exploring the knowledge and capacities involved in using curriculum resources well. He used the term *pedagogical design capacity* to refer to an individual teacher's ability to perceive and mobilize curricular resources in order to design instruction. Although Brown did not offer further conceptualization of this construct, other researchers have begun to explore this question, looking distinctly at perceiving, mobilizing, and considering the understandings, skills, and inclinations these activities require.

Perceiving includes reading, making meaning of, and evaluating the contents of curriculum resources to understand what is available mathematically and pedagogically and deciding whether and how to use them. We know from research on print materials that teachers read different components of the resource for different reasons (Remillard & Bryans, 2004; Sherin & Drake, 2009; Stein & Kaufmann, 2010). We also know that perceiving the mathematical purpose embedded in instructional representations, tasks, and sequences is not straightforward and requires a certain type of content-specific curriculum knowledge (Choppin, 2011; Kim & Remillard, under review). The opaque and layered nature of digital resources is likely to make reading the mathematics even more challenging.

Mobilizing involves leveraging the guidance and tools in instructional resources into plans for instruction, and enacting and adapting those plans in real time. It is not surprising that mathematical knowledge comes into play in this work (Hill & Charalambous, 2012), but others have suggested it is not sufficient. Stein and Kaufmann (2010) found that teachers who read the explanations of the key mathematical ideas in their teacher's guides, regardless of their mathematical knowledge for teaching (MKT),[7] were more likely to maintain the cognitive demand of the tasks during enactment. Choppin (2011) found that understanding of the instructional sequences in curriculum guides influenced the strength of their adaptations. The flexible and nonlinear characteristics of digital resources, the many possible paths through the domain, together with the fact that they are often packaged in discrete chunks, increases the need for teachers to see beyond a single task. Teachers must be able to discern or create an instructional pathway through a mathematical domain using a range of available tools. This type of mapping requires a great deal of curricular knowledge, which has, traditionally, been offloaded to curriculum authors or mathematics supervisors. Using digital resources increases the demand on teachers to adopt a broader curricular vision, a perspective many teachers have not had opportunities to develop. Research is needed to understand this knowledge, how it is developed, and what it looks like in practice.

Can Digital Tools Assist Teachers in Re-Sourcing?

One final question the field must continue to explore is how resources might be designed to guide and assist teachers in navigating the digital world. Some researchers and designers have proposed that resources might be designed to be educative for the teachers using them as well as for students (Davis & Krajcik, 2005) and have offered a process for designing educative materials (Davis et al., 2014). Although some may be under the impression that digital resources reduce the demands on teachers, the characteristics of digital resources discussed previously suggest that they place even greater and particular types of demands on teachers. The question that follows is: Could educative features be built into digital resources to assist teachers in meeting these demands? For example, can these resources include frameworks or decision-making structures that guide teachers in making navigational decisions? Questions like these can be explored by those designing digital curriculum resources and by researchers studying their use by teachers.

ACKNOWLEDGMENTS

I would like to acknowledge the many colleagues and graduate students who have been part of the research I have done or have contributed to my thinking in critical ways. Among these are members of the ICUBiT (Improving Curriculum Use for Better Teaching) research team: Ok-Kyeong Kim, Napthalin Atanga, Shari Ciganic, Nina Hoe, Luke Reinke, Dustin

Smith, Joshua Taton, and Hendrik Van Steenbrugge. (ICUBiT is funded by the National Science Foundation, under Grant Nos. 0918141 and 0918126. Any opinions, or recommendations expressed in this chapter are those of the authors and do not necessarily reflect the views of the National Science Foundation.) I have also benefited tremendously from the CSMC community for the past 10 years.

NOTES

1. See Pepin, Gueudet, & Trouche (2013). Kenneth Ruthven (2016) elaborates on this term.
2. For more on this perspective, see Remillard (2005, 2012).
3. See the chapter by Ruthven (2016) for more detail.
4. See chapter by Yerushalmy (2016). Her Visual Math eBook provides such an example.
5. Dorsey (2016)
6. See the chapter by Ruthven (2016) for a helpful characterization of different approaches of drawing, using, and coordinating different online resources.
7. Mathematical Knowledge for Teaching is a construct first introduced by Thompson and Thompson (1996); Hill, Schilling, and Ball (2004) have further conceptualized this domain of knowledge and developed measures to assess it.

REFERENCES

Brown, M. W. (2009). The teacher–tool relationship: Theorizing the design and use of curriculum materials. In J. T. Remillard, B. A. Herbel-Eisenmann, & G. M. Lloyd, (Eds.), *Mathematics teachers at work: Connecting curriculum materials and classroom instruction* (pp. 17–36). New York, NY: Routledge.

Choppin, J. (2011). Learned adaptations: Teachers' understanding and use of curriculum resources. *Journal of Mathematics Teacher Education, 14,* 331–353.

Davis, E. A., & Krajcik, J. S. (2005). Designing educative curriculum materials to promote teacher learning. *Educational Researcher, 34*(3), 3–14.

Davis, E. A., Palincsar, A. S., Arias A. M., Bismack, A. S., Marulis, L. M., & Iwashyna, S. K. (2014). Designing educative curriculum materials: A theoretically and empirically driven process. *Harvard Educational Review, 84*(1), 24-52.

Dewey, J. (1897). My pedagogic creed. *The School Journal, 54*(3), 77–80.

Dorsey, C. (2016). Deeply digital STEM learning. In M. Bates & Z. Usiskin (Eds.), *Digital curricula in school mathematics* (pp. 285–296). Charlotte, NC: Information Age.

Gueudet, G., & Trouche, L. (2009). Towards new documentation systems for mathematics teachers. *Educational Studies in Mathematics, 71*(3), 199–218.

Herbert, M. (1984, April). *Comprehensive school mathematics program: Final evaluation report.* Aurora, CO: Mid-continent Regional Educational Laboratory.

Hill, H. C., & Charalambous, C. Y. (2012). Teacher knowledge, curriculum materials, and quality of instruction: Lessons learned and open issues. *Journal of Curriculum Studies, 44*(4), 559–576.

Hill, H., Schilling, S., & Ball, D. L. (2004). Developing measures of teachers' mathematics knowledge for teaching. *The Elementary School Journal, 105*(1), 11–30.

Kim, O.K., & Remillard, J. T. (under review). *Knowledge of curriculum embedded mathematics: A critical domain of teacher knowledge.*

Moursund, D. (2016). Mathematics education is at a major turning point. In M. Bates & Z. Usiskin (Eds.), *Digital curricula in school mathematics* (pp. 271–284). Charlotte, NC: Information Age.

Pepin, B., Gueudet, G., & Trouche, L. (2013). Investigating textbooks as crucial interfaces between culture, policy, and teacher curricular practice: Two contrasted case studies in France and Norway. *ZDM—The International Journal on Mathematics Education, 45*(5), 685–698.

Remillard, J. T. (2005). Examining key concepts of research on teachers' use of mathematics curricula. *Review of Educational Research, 75*(2), 211–246.

Remillard, J. T. (2012). Modes of engagement: Understanding teachers' transactions with mathematics curriculum resources. In G. Gueudet, B. Pepin, & L. Trouche (Eds.), *From text to "lived" resources: Mathematics curriculum materials and teacher development* (pp. 105–122). New York, NY: Springer.

Remillard, J. T., & Bryans, M. B. (2004). Teachers' orientations toward mathematics curriculum materials: Implications for teacher learning. *Journal for Research in Mathematics Education, 35*, 352–288.

Ruthven, K. (2016). The re-sourcing movement in mathematics teaching: Some European initiatives. In M. Bates & Z. Usiskin (Eds.), *Digital curricula in school mathematics* (pp. 75–86). Charlotte, NC: Information Age.

Sherin, M. G. & Drake, C. (2009). Curriculum strategy framework: Investigating patterns in teachers' use of a reform-based elementary mathematics curriculum. *Journal of Curriculum Studies, 41*(4), 467–500.

Stein, M. K., & Kaufman, J. H. (2010). Selecting and supporting the use of mathematics curricula at scale. *American Educational Research Journal, 47*(30), 663–693.

Thompson, A. G., & Thompson, P. W. (1996). Talking about rates conceptually, Part II: Mathematical knowledge for teaching. *Journal for Research in Mathematics Education, 27*(1), 2–24.

Usiskin, Z. (2010). The current state of the school mathematics curriculum. In B. Reys, R. Reys, & R. Rubenstein (Eds.), *Mathematics curriculum: Issues, trends, and future direction, 72nd yearbook* (pp. 25–39). Reston, VA: National Council of Teachers of Mathematics.

Usiskin, Z. (2013). Studying textbooks in an information age—A United States perspective. *ZDM: International Journal on Mathematics Education, 45*(5), 713–723.

Vygotsky, L. S. (1978). *Mind in society.* Cambridge, MA: Harvard University Press.

Yerushalmy, M. (this volume). Inquiry curriculum and e-textbooks: Technological changes that challenge the representation of mathematics pedagogy. In M. Bates & Z. Usiskin (Eds.), *Digital curricula in school mathematics* (pp. 87–106). Charlotte, NC: Information Age.

CHAPTER 14

NEW STARTING POINTS FOR NUMBER SENSE USING *TOUCHCOUNTS*

Nathalie Sinclair

INTRODUCTION

Learning to count in a contemporary world is as basic as learning to walk and talk. However, it is known that young children often experience difficulty in creating a one-to-one correspondence between the counted objects and assigning the number attributed to the last counted object as an enumerator of the total. Furthermore, initial experiences with arithmetic operations may present a challenge for learners, especially when the operations are approached by means of direct modeling (Coles, 2014). How can technology assist with these challenges? One vision of technology is that it facilitates learning, often by offering added pragmatic value that makes it easier or more efficient to carry out tasks (like graphing a function on a graphing calculator or generating a counterexample in a dynamic geometry environment) (Artigue, 2002). Another vision of technology is that it changes the nature of learning as well as the nature of mathematics itself (Papert, 1980). It is this latter view that has motivated the development of

TouchCounts[1] (Sinclair & Jackiw, 2011), a multitouch app designed to enable finger-based interactions with ordinal and cardinal aspects of numbers.

There has been extensive research on the development of early number sense, much of which is based on learning environments that do not include digital technologies. The influential work of Gelman and Meck (1983), for example, proposes five counting principles that culminate in the cardinal principle, mastered around the age of 8, which asserts that the last number word of an array of counted items represents the numerosity of this set of items. Indeed, cardinality is widely viewed by researchers and curriculum developers as the main goal of early number development. Much of the pedagogy associated with this approach involves linking number symbols to collections of objects, such as having children count sets of objects. The development of the cardinality principle is seen as essential for working with operations. Related to this, instead of focusing on the definition of the concept of cardinality, researchers such as Vergnaud (2008) have reformulated cardinality in terms of what cardinal numbers *do*, that is, in more functional terms: cardinal numbers can be operated on. The importance of cardinality has also promoted increased attention to subitizing (see Clements, 1999).

An alternate approach involves prioritising ordinality, which can be seen, for example, in the pedagogical approach of Gattegno (1974). This approach conceptualizes numbers as relations (bigger, smaller, equal) and emphasizes work with symbols (rather than with physical objects). Recent neuroscientific work has suggested that ordinality may indeed be very important in children's learning, particularly when they begin to work with arithmetic operations (Lyons & Beilock, 2011, 2013; Lyons, Price, Vaessen, Blomert, & Ansari, 2014).

In some respects, *TouchCounts* can be seen as changing both the learning of mathematics and the nature of mathematics itself in that it provides forms of feedback and engagement that affect the ways children can count and operate. For example, children can operate on numbers by pinching them together, long before they can be expected to have developed a sense of cardinal numbers according to traditional research. In addition, their counting activities permanently feature numerical symbols in addition to the visual items that are being counted and the auditory word names that accompany each item. These activities immerse children into a kind of "Mathland" (to use Papert's term) where they might see (and also hear) numbers that are much bigger than they have previously encountered. In the next section, I provide a brief overview of *TouchCounts*. I then describe excerpts from research videos that highlight some of the more significant changes that it offers, particularly in comparison to non-digital environments.

A MULTI-TOUCH APPLICATION FOR COUNTING AND OPERATING

The multi-touch device is a novel technological affordance in mathematics education. Through its direct interaction, it offers opportunities for mathematical expressivity by enabling children to produce and transform screen objects with fingers and gestures, instead of engaging and operating through a keyboard or mouse. This makes it highly accessible, but also opens the way for new, tangible forms of mathematical communication (Jackiw, 2013).

Unlike many "educational games" that can be found for the iPad, *TouchCounts* is open-ended and exploratory, rather than practice- and level-driven—it follows in the tradition of constructionist and expressive technologies in mathematics education (Noss & Hoyles, 1996; Papert, 1980) and supports the development of number by offering modes of interaction with objects that involve fingers and hands. Specifically, it aims both (a) to engage one-to-one correspondence by allowing every finger touch to summon a new sequentially-numbered object into existence, one whose presence is both spoken aloud and symbolically labelled, and (b) to enable gesture-based summing and partitioning, by means of pushing objects together and pulling them apart in ways that expose very young children to arithmetic operations. With these new affordances, however, come new questions related to design decisions—such as "What touch-based actions on the screen might better support and enable mathematical activity?"—as well as questions related to the development of number and how this particular technology may shape current curricular trajectories and, in the process, potentially disrupt them.

Currently, there are two sub-applications in *TouchCounts*, one for enumerating and the other for operating. After describing each of the two worlds, I present and analyze a series of tasks, where the first set are to be used in the Enumerating World and the second set in the Operating World.

The Enumerating World

In this world, a user taps her fingers on the screen to summon numbered objects (yellow discs). The first tap produces a disc containing the numeral "1." Subsequent taps produce successively numbered discs. As each tap summons a new numbered disc, *TouchCounts* audibly speaks the number word for its number ("one," "two,"..., if the language is set to English). Fingers can be placed on the screen one at a time or simultaneously. With five successive taps, for instance, five discs (numbered "1" to "5") appear sequentially on the screen, which are counted aloud one by one (see Figure 14.1a). However, if the user places two fingers on the screen simultaneously, two consecutively numbered discs appear at the same time (Figure 14.1b), but only the higher-numbered one is named aloud ("two," if these are the first two taps). One small instance of opportunity lies in a new sense of the times two table: the number of "times" two fingers simultaneously touch the screen. The entire "world" can be reset to clear all numbered discs and return the "count" of the next summoned disc to one. The discs always arrive in order, with their symbolic names imprinted upon them.

From an adult perspective, the number of taps (whether made sequentially or simultaneously) is also the number of discs on the screen, a fact that can tacitly reinforce the cardinality principle, since the last number "counted" (spoken aloud by *TouchCounts*) is exactly "how many" numbered discs there are to be seen. In traditional research in the area of early counting, it is well-documented that, even after children have counted a set of things (up to five, say), when they are asked "how many" objects are in that set, they will often count the objects again (Baroody & Wilkins, 1999). The "how many?" question seems to provoke a routine of sequential counting.

In *TouchCounts*' Enumerating World, however, the child is engaged in a somewhat different practice—rather than counting a *given* set, she is actively *producing* a set with her finger(s) (perhaps aiming at a pre-given total) and the elements of that set seem to count themselves (both aurally and symbolically) as they are summoned into existence. One distinction that *TouchCounts* makes is that, orally, each number word in succession replaces (and eradicates) the previous one. At the end of the spoken count, no trace is left of what has been said. On the screen, however, each action leaves a visual trace, in the form of (one or more) numeral-bearing discs, of what has once been summoned into being.

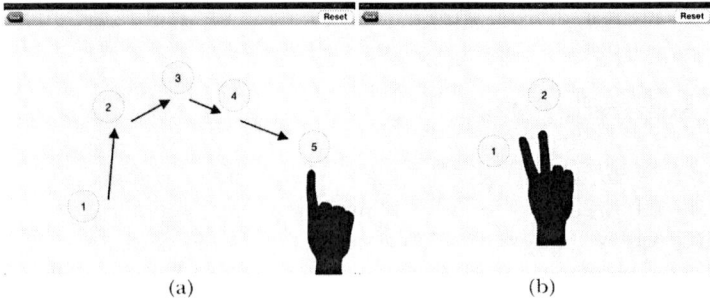

Figure 14.1 (a) Five sequential taps—"one, two, three, four, five" is said (the arrows are only to indicate the sequence; they are not shown on the screen); (b) A simultaneous two-finger tap—only "two" is said (both discs appear simultaneously).

If the "gravity" option for this world is turned on, then as long as the learner's finger remains pressed to the screen, the numbered object holds its position beneath her fingertip. But as soon as she "lets go" (by lifting that finger), the numbered object falls and then disappears "off" the bottom of the screen, as if captured by some virtual gravity. With "gravity" comes the option of a "shelf," a horizontal line across the screen. If a user releases her numbered object above the shelf, it falls only to the shelf and comes to rest there, visibly and permanently on the screen, rather than vanishing out of sight "below." (Figure 14.2 depicts a situation in which there have been four taps below the shelf—these numbered objects were falling—and then a disc labelled "5" was placed above the shelf by tapping above it). Since each time a finger is placed on the screen a new numbered disc is created underneath the finger and, once released by lifting the finger it begins to fall, one cannot "catch" or reposition an existing numbered object by re-tapping it. This is not a conventional "dragging" world.

Discs dropping away under "gravity" mirror the way spoken language fades rapidly over time, with no trace left—the impermanence of speech. Also, with discs disappearing, any sense of cardinality goes too: in the absence of the presence of "1," the disc labeled "2" is simply the second one to have been summoned by the tapping. So, the Enumerating World with "gravity" enabled (it is an option) is almost entirely an ordinal one, with the shelf acting as a form of visible memory.

One of the characteristics of *TouchCounts*, then, is that the computer handles the counting (the iPad is the one who announces and manages the arrival of various figures onto the ritual scene). The design intent is to help move young users towards transitive counting (or counting objects), even though the general setting provides a mix of cardinal and ordinal elements.

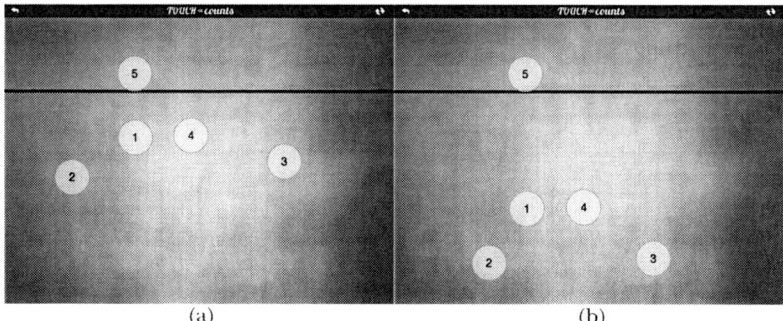

Figure 14.2 (a) After four sequential taps below the shelf and a fifth tap above the shelf; (b) The discs below the shelf fall down.

The Operating World

Whilst tapping on the screen in the Enumerating World creates sequentially numbered objects, tapping on the screen in the Operating World creates autonomous numbered sets, which we refer to as *herds*. The user's creation choreography starts by placing one or several fingers on the screen, which immediately creates a large disc that encompasses all the fingers and includes a numeral corresponding to the combined number of fingers touching the screen. At the same time, every one of the fingers in contact with the screen creates its own much smaller (and unnumbered) disc, centered on each fingertip. When the fingers are lifted off the screen, the numeral is spoken aloud and the smaller discs are then lassoed into a herd and arranged regularly around the inner circumference of the big disc (Figure 14.3a shows herds of 3 and 4). The small discs all move in either a clockwise or counter-clockwise direction to emphasize they are to be seen as one unit. Herds of size one wander around the screen in order to make them more difficult to place one's finger on, in order to encourage children to operate with herds that are greater than one in number.

Unlike in the Enumerating World, herds can be interactively touched, either to move them around on the screen or to operate upon them. After two or more such arrangements have been produced (as in Figure 14.3b) they can either be pinched together (addition) or "unpinched" (subtraction or partition). In the case of pinching together, two fingers are required—one on each herd—to make the herds merge. Dynamically, they then become one herd that contains the "digital" counters from each previous herd, thus adding them together. The new herd is labeled with the associated numeral of the sum (Figure 14.3c), which *TouchCounts* announces aloud. Moreover, the new herd keeps a trace of the previous herds, which can be seen by means of the differentiated colors of the individual small discs. Multiple herds can be pinched together simultaneously. The pinching gesture is entirely symmetric, both with respect to the pinching fingers and with respect to the herds, so that adding does not have the kind of order implied by the

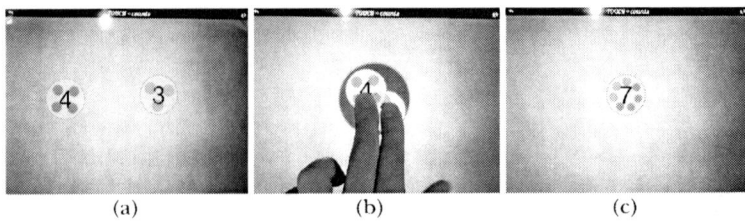

(a)   (b)   (c)

Figure 14.3 (a) The herds, (b) Pinching two herds together, (c) The sum of two herds.

directionality of verbal or written expressions such as "two plus three" or "2 + 3."

An inverse pinch gesture ("unpinching") can be made in order to decompose a given herd into two herds (Figure 14.4a). The gesture can be described either as "separating," which supports the idea of partitioning, or as "taking out" or "removing," which supports the idea of subtracting. In both cases, two fingers are placed in the herd—while one stays put, the second swipes out of the herd. This distinction of roles between the two fingers supports the needed directionality of subtraction. The further the swipe travels, the more will be taken out from the starting herd (and of course, at the extreme, everything can be taken out of the starting herd) (Figure 14.4b). When the swiping finger is lifted, two new herds are formed and *TouchCounts* announces the number that has been taken out (Figure 14.4c). In the extreme case (where everything is removed), a new herd is formed under the finger that has swiped, while in the location of the previous herd the numeral "0" appears briefly but then fades away.

Children can create and merge or split herds without planning to or even without knowing that they are also adding or subtracting. Indeed, the pinching gesture draws on one of the four grounding metaphors for addition, that of *object collection* (see Lakoff & Núñez, 2000). Both adding, and either subtracting or partitioning, offer children the *action* of operating without necessarily requiring them to calculate the result. Unlike with the calculator, which can also perform addition and subtraction, *TouchCounts* first requires the production of herds that will be labeled by a numeral (indicating "how many" are in the herd) and then enacts the gathering/splitting mechanisms in which the two herds join or separate, both visually and temporally.

Children can pinch two herds together or split a herd apart relatively easily (though some children find it challenging, at least initially, to place their fingers right on the herds and often produce new herds). They can do this, obviously, without knowing what the sum or difference will be, without knowing that the transformation occurring reflects the operation of

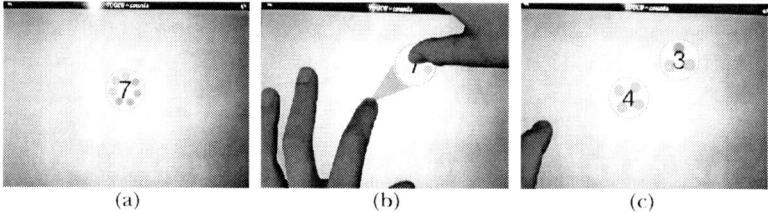

Figure 14.4 (a) An initial herd of 7, (b) Left finger (in this instance) swiping outside the herd, (c) Resulting separation of 7 into herds of 4 and 3: TouchCounts announces "four."

addition or subtraction and, most importantly, without thinking of those herds as cardinal numbers. In this sense, *TouchCounts* invites the children into a gesture-mediated form of operations. While a teacher might introduce the word "adding" to describe the process, neither that word nor its symbolic counterpart (e.g., "+") appears on the screen: the iPad is silent while pinching or unpinching occurs. As such, language such as "making" or "putting together" or "joining" can all be used to accompany the action (the emerging rite) of pinching discs together.

HIGHLIGHTS FROM RESEARCH

I present here a selection of excerpts, many of which have been described in more detail elsewhere (see Sinclair & Heyd-Metzuyanim, 2014; Sinclair & Pimm, 2014). My focus here is on the particular features of children's interactions with *TouchCounts* that may have important curriculum considerations, either in terms of how *TouchCounts* might be integrated into current curricular expectations or in terms of how it might evoke the potential for different kinds of learning paths. The first two episodes are set in the same day-care facility in which my research team set up a desk in one corner of the room and children (aged three to five) were allowed to come and go as they wish (as per the culture of this day-care). Most of the time, three or four children were involved at a time. I allowed the children to explore, but also proposed certain tasks depending on the children's particular interests as well as their experience of using *TouchCounts*. The sessions lasted about one hour, but the children tended to spend between five and fifteen minutes at a time. The third and fourth excerpts come from a different setting involving individual interactions between me and kindergarten children (aged 5), who were pulled out of their regular classrooms for my research.

I Made Eighteen

This episode involves four children all aged either three or four. I asked one of the children, Rodrigo (who was four and had participated in previous sessions), to "make five," which he did by tapping with his left index finger along an imaginary horizontal line from left to right, stopping exactly after *TouchCounts* said "five." While he was doing this, two of the girls who were crowded around the iPad started leaning back and one looked away. A third girl (Grace), aged three, who had not played with *TouchCounts* before and who had been watching carefully, said, "I want to do it now." I said "Okay" and placed the iPad in front of her. Rodrigo was still watching as Grace started tapping but not saying anything (Figure 14.5a). She started on the left

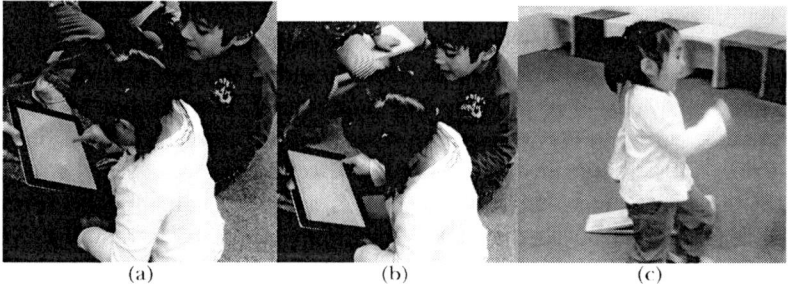

Figure 14.5 (a) Grace tapping on the screen, (b) Grace tapping on the third row, (c) Grace strutting around the classroom shouting, "I made eighteen!"

and then tapped towards the right of the screen along a horizontal, until she reached the edge of the screen at "six" (with *TouchCounts* counting along). Instead of stopping there, she tapped underneath her numbered disc 1 (thus mirroring writing conventions) and then continued tapping toward the right until she reached the screen edge again, at "twelve." She then placed the numbered disc 13 under the one showing "7," close to the bottom of the screen, and she tapped until she got to the right edge a third time at "eighteen" (Figure 14.5b) and then took her finger away from the screen.

As alluded to previously, the classic exchange in this area of early number starts with an emphasis on cardinality, posing the question: "How many?" In the initial prompts here, I started with this question as well, even though it might not have been the most appropriate one. This point provides a strong example of how technology can determine "the right question to ask," as well as indicating how some of the apparent "facts" about early number are conditioned by the very questions used to ascertain them.

> **I:** How many did you make?
> **Rodrigo:** (*Answering for her*) Sixteen, seventeen, s'teen. (*The sounds before "teen" are hard to hear, but the words are said in a rhythmic manner as he points in turn to discs on the screen.*)
> **Grace:** (*Looking up at the adult, smiling*) Eighteen! (*Pointing to the numbered disc 18 at the bottom right of the screen*)
> **I:** You made eighteen. (*Presses Reset*) Now it's my turn.
> **Grace:** (*Gets up and walks away from the group into the middle of the room*) I made eighteen! Eighteen, eighteen, eighteen, eighteen, eighteen, eighteen, eighteen. (*Marching around the room as she calls out*—see Figure 14.5c).

This episode is interesting for several reasons, not least Grace's excitement at having made eighteen, which was not given as a task by me, but which emerged from her own exploration. Her pride in having made what she did (which is expressed in her excited "I made eighteen") underscores the fact that it was indeed *she* who *made* it, even though it was not she who initially proclaimed it (the iPad actually said "eighteen"). Further, her excitement shows how a seemingly arbitrary number (not the more common ones usually discussed in educational settings like five, ten, or even twenty) can be so interesting to Grace—she may not have frequently dealt with this number before and it possibly strikes her as excitingly big. The Mathland of *TouchCounts* thus immersed her in numbers outside of the range of numbers (1 through 10) that most preschool standards mandate.

Next, the mobility and flexibility of using the iPad (unlike using a computer) makes Grace's coming to the iPad and then her happy strut away from it flow naturally within the activity of the room, which is consonant with the philosophy of this play-based day-care. Lastly, eighteen had become a point of arrival, rather than simply another number word in the on-going stream of the count. What is even more surprising, given her very young age, is the fact that she immediately says "Eighteen!" rather than counting all the numbered discs again. Perhaps Rodrigo's mumbling about "teens" oriented her attention to what she had just heard spoken aloud, which dissuaded her from counting them all up. Perhaps because she had made them with her finger, without having to keep count of them herself, she could attend more to the importance of the last spoken number and perhaps associate it with the quantity of discs on the screen.

This brief example shows, at the very least, that *TouchCounts* muddies simple assertions about the cardinality principle and children's ability to work with cardinal numbers. I am not saying that Grace has learned about cardinality in this brief experience. Rather, I am interested in how the tactile, auditory, and symbolic richness of her counting activity changes the accessibility of cardinality in that she can both see the numeral 18 and recall the word "eighteen" having been said aloud. The "how many?" question may thus be of limited interest, both to her and me. A more appropriate question might have been, "What did you make?" Follow-up questions might include "What will come next?" In this example, both ordinality and cardinality seem to be in play, with sequential finger tapping producing the rhythmic enumeration of numbers with their correctly ordered symbols and the count functioning as a marker of the last number produced.

Inverse Subitizing

In subitizing tasks, students must determine quickly the number of objects in an array, which they then either say or type using a keyboard. One of the *TouchCounts* tasks is "inverse subitizing," where children were asked to produce a target number (say, 5) by placing several fingers all-at-once (rather than sequentially) on the screen. Unlike subitizing, which involves producing a spoken or alphanumeric action based on a visual prompt, inverse subitizing requires that the children produce an action (five fingers placed on the screen simultaneously) based on an oral prompt ("make five all-at-once"). Unlike conventional subitizing tasks, which rarely extend beyond five, inverse subitizing with *TouchCounts* has no upper limit, in the sense that a child may use all her fingers to make ten/10, but she can also work collaboratively with other children to produce even higher targets.

This example involves inverse subitizing in the Enumerating World (in the no-gravity mode). A group of students (3, 4, and 5 years old) have all successfully made four "all-at-once" (by placing four fingers on the screen simultaneously). I ask Owen whether he would like to try to make "seven all at once." Owen looks at his hand, then turns them around so that they are palm up. He turns one hand and uses it to count off five fingers on the other hand, counting them aloud one by one. He then turns the counting hand and lifts two fingers, one by one, counting them as "six" and "seven." He then turns both hands over so that the fingers are pointing down, and touches them to the screen. *TouchCounts* says "eight" (he had mistakenly touched the screen with another part of his hand as well as with the 7 fingers). I ask if he wants to try again and he nods. He then immediately stretches out the same seven fingers he had before and places them on the screen, which makes *TouchCounts* say "seven."

Owen had thus quickly developed a very specific gesture for making 7, which meant that he did not have to count 7 out one by one on the second try. Instead, he could immediately use the "septet" he had previously created as a cardinal number. Indeed, when asked later on to add 7 and 2, Owen began with the septet and then lifted two more fingers. This kind of "inverse subitizing" work may thus more directly support the development of arithmetic, especially because it keeps the number thinking in the hands, rather than focusing on the visual assessment of numerosity associated with typical subitizing tasks.

This quick shift from a counting out of the fingers to an immediate production of a set of numbers is something that has occurred over and over again in our research studies. For young children especially, it seems closely tied to the development of finger gnosis, which involves much inspection of the actual fingers, but also the reciprocal touch of the screen and the fingers.

What Comes After 6?

Katy, a five-year old kindergarten student, had worked for several minutes on the task of placing "just five" on the shelf. This involves someone tapping four times below the shelf and then once above it. After successfully doing so (on her fourth attempt), the teacher asked her, "Could you just put five and then ten up there?" (indicating the shelf). Katy started creating discs below the shelf. She succeeded in putting 5 above the shelf, then tapped below the shelf, producing six/6, and looked up at me. She continued tapping below the shelf to get "seven," and "eight," and then "nine," each time looking up at me and each time telling her that she did not want to see those numbers. I pressed Reset and asked Katy to pretend that five and ten were her friends whom she wanted to stay on the shelf. Katy produced 1, 2, 3, and 4 below the shelf, then 5 above the shelf and then 6 below it again.

> **Katy:** What kind of number is going to come after? (*In other words, friend or not friend, so to go above or below the shelf*).
> **I:** After six? What do you think?
> **Katy:** Don't know. One, two, three, four, five, six, seven! (*Says this sequence very fast; taps once more below the shelf and then looks up*).
> **iPad:** Seven.
> **Katy:** Eight. Does he go there?
> **I:** He's not your friend, just ten.

Katy tapped below the shelf, the iPad said "eight," and then Katy asked "Nine?" and looked at me. I said that nine was "not your friend." She asked "Is nine going to come after?", and when I responded, "You just did eight, what do you think?", she began to count aloud starting from one once more. She counted fast orally, taking over for the iPad, but skipping some numbers, and went all the way to nineteen, made a grimace and said "No!" She then tapped below the shelf (for 9) and looked at me.

> **I:** (*Looking at Katy*) And after nine?
> **Katy:** (*Taps below the shelf*)
> **iPad:** Ten.
> **I:** Oh! Ten was our friend! We wanted to have him.
> **Katy:** (*Looks back at screen*) One and an oh. (*Peers closely at the screen, smiles, and raises her index finger*).

After this moment of disappointment, Katy started over again. She tapped four times below the shelf, looked up at me and then tapped the fifth time above the shelf. Her subsequent tapping again below the shelf produced 6, 7, and 8 below the shelf and then she looked up at me, eventually deciding to tap once more below the shelf, which produced the 9 there. She then paused, looked up at me, and was about to place her finger *on* the shelf when I said that she should make sure to place it *above* the shelf if she

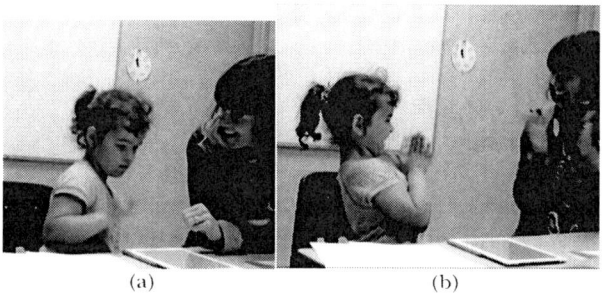

Figure 14.6 (a/b) "You got both your friends!"

wanted it to stay. Katy tapped above the shelf, producing 10. I exclaimed happily "Yeah! Good job! You got both your friends!" (Figure 14.6a) and started clapping. Katy starting clapping too (Figure 14.6b).

It had taken Katy some time to learn how to put "just five" on the shelf, but as she moved to the harder task (of getting 10 there), she could anticipate exactly when to move her finger above the shelf—that is, right after she had produced 9. She learned to do this thanks in part to the combined visual and auditory feedback of *TouchCounts*, both available only temporarily, since the numbered discs fell away and the number words were each said aloud only once. After a while, five (and later, ten) was no longer solely the end-point of a process of counting things, but had become the one that follows four (nine), and could stand alone on the shelf without its predecessors being present.

Her explicit question ("What comes after?") shows that Katy was deciding where to tap based on what comes after the one that was just created (and said aloud)—ordinality. Katy did not seem sure what came after six or eight and had to count up with a running start from one. (Unlike the suggestion about eighteen with Grace, Katy has to re-immerse the words in the temporal stream of the count to be sure of the next one.) She was evidently not prepared for the arrival of the disc bearing the number symbol 10, but noticed as it fell off the screen, saying "One and a oh." On the next try, Katy did not need to think about what follows six and eight. While it may be that she had remembered the successors of seven and eight, it is also possible that Katy had learned to make a rhythmic tapping (four below, one above).

With this task, Katy was exploring the neighborhood between one and ten, rehearsing her counting, but also attending to which words/symbols precede and follow a given pre-announced target. Her rhythmic way of tapping seems to give a sense of five as four taps below followed by one above (a rhythm that echoes tallying, where after four vertical strokes have been recorded a fifth line is drawn through the other four to make five), which serves not only to set five apart, but also to group the four falling below as a whole (the not-fives), one which relates to its cardinality.

From Skip Counting to Multiples

In this last example, I focus on another self-initiated exploration undertaken by a five-year-old girl named Chloe. Chloe had been invited into the same task described in the example of Katy and had quickly succeeded in placing both 5 and 10 on the shelf. I thus moved to the Operating World and asked her to place several fingers on the screen at once. I then showed her how to pinch two herds together and let her try doing so. After she pinched together a few herds, I asked Chloe if she could "make a five." She did that very easily, both by tapping all her five fingers together at once, and then by tapping with her index finger five times and then trying to pinch the separate "one" herds into a herd of five.

Then I asked her whether she could make seven. She did this initially by pinching a herd of five and two together. But when I asked her to make seven in a different way, it took her several minutes, with some experimenting, before arriving at a solution of pinching a four and a three together. She was very proud of her accomplishment, but seemed tired as well. After stretching and resting for a few seconds, Chloe leaned towards the iPad, uttered an "Umm..." and paused for a few seconds. I asked, "What are you thinking about?" and Chloe responded, "I'm thinking about making, adding ten and ten. I wonder what that makes." I showed her a technique that could make it easier for her to do that. I put my five fingers on the screen, then added five "ones" with my index finger already in the big "five" circle. This added the numbers straight into the herd without the need to pinch them together. I let Chloe try this.

> **Chloe:** (*Places right hand on screen*; Figure 14.7a)
>   **I:** Yeah, so you're at five now.
> **Chloe:** [...] make it huge. (*Looking down at her fingers and spreading them out so that the lasso becomes bigger*)
>   **I:** Now you have six actually.
> **Chloe:** (*Uses left hand index finger to tap four times*)
>   **I:** Let go.
> **Chloe:** (*Lifts her hand from the screen*)
>   **iPad:** Ten.
> **Chloe:** (*Places right hand on screen then taps five times with left index finger and then lifts her hand from the screen*)
>   **iPad:** Ten.
>   **I:** Very nice.

Figure 14.7 (a) Chloe making five, (b) Pinching the two groups of ten together, (c) Sharing her discovery with the adult.

> **Chloe:** (*Pinches the two herds together with her right hand and then lifts her hand from the screen;* Figure 14.7b) Twenty?
> **I:** Twenty.
> **Chloe:** (*Looking up*) That's why they say five, ten, (*short pause*) fifteen, twenty. (*Looks at me*; Figure 14.7c)

Chloe continued this way, making herds of ten and pinching them to her ever-growing herd of twenty, thirty, forty, ... ninety, until she reached a hundred. She seems to have known that she would be able to use the pinching gesture to create a large number like one hundred. However, as evident in her questioning tone when the pinch gesture created a herd of twenty, Chloe had not anticipated this result. Once she saw this new herd, Chloe looked up, with a smile, and uttered a statement that connected skip counting by five (which is very similar to the intransitive counting sequence) to the successive adding of five. In other words, she was linking something she had heard ("*they* say") with the action she had just employed of adding ten and ten to get twenty.

DISCUSSION

In terms of curriculum integration, the current emphasis on cardinality may make it easier to integrate some aspects (or activities) of *TouchCounts* in current practices, including the improvement of finger gnosis and subitizing. However, the possibility of being able to create very big numbers (both in the Enumerating and the Operating Worlds) may not be welcomed by teachers and researchers who focus on the importance of developing a sense of smaller quantities. To be sure, while the children in the research studies we have conducted are able to create very large numbers, they may not have a sense of the quantities these numbers are associated with. However, they are able to think of these numbers relationally; for example, they make statements such as "two hundred and four is after one hundred" and "you need another hundred to get to two hundred" (from one hundred).

Such statements arise from experiences of counting very high and noticing that two hundred and four is said and seen after one hundred, which is different from assessing relative size based on place value, for example.

Similarly, some researchers and teachers have requested a setting that enables the numerical symbols to be hidden. This is in line with a cardinally focused sequence of instruction in which numbers are linked to physical objects (and where symbols do not initially appear). While I have noticed that young children do not focus on the symbols in their first interactions, they definitely notice them eventually (like Katy noticing the 10). *TouchCounts* might be easier to integrate, therefore, in an ordinally focused sequence of instruction where the numerical symbols play an earlier and more integral role. Finally, given that iPads are now pervasive in many homes, and that children will be using *TouchCounts* with parents and other care-givers, there are interesting questions related to how the app integrates with non-school based environments. Thus far, we have noticed that parents are not always sure how to interact with their children, and what kinds of questions to pose or challenges to offer. This contrasts with most other Apps, where the goals are clear and the parents not needed. One of the research questions I am currently pursuing is how to best support these home-based experiences, and how such support may differ from the support that might be appropriate for teachers.

NOTE

1. *TouchCounts* is available for free at the App Store. Readers are also encouraged to visit the *TouchCounts* Youtube channel, which contains several videos exemplifying some of the activities described in this chapter.

REFERENCES

Artigue, M. (2002). Learning mathematics in a CAS environment: The genesis of a reflection about instrumentation and the dialectics between technical and conceptual work. *International Journal of Computers for Mathematical Learning, 7*(3), 245–274.

Baroody, A., & Wilkins, J. (1999). The development of informal counting, number, and arithmetic skills and concepts. In J. Copley (Ed.), *Mathematics in the early years* (pp. 48–65). Reston, VA: National Council of Teachers of Mathematics.

Clements, D. H. (1999). Subitizing: What is it? Why teach it? *Teaching Children Mathematics, 5*(7), 400–405.

Coles, A. (2014). Ordinality, neuro-science and the early learning of number. In C. Nichol, S. Oesterle, P. Liljedahl, & D. Allen (Eds.), *Proceedings of the joint PME*

38 and PME–NA 36 conference (Vol. 2, pp. 329–336). Vancouver, BC: Psychology of Mathematics Education.

Gattegno, C. (1974). *The common sense of teaching mathematics*. New York, NY: Educational Solutions.

Gelman, R., & Meck, E. (1983). Preschoolers' counting: Principles before skill. *Cognition, 13*(3), 343–359.

Jackiw, N. (2013). Touch and multitouch in dynamic geometry: Sketchpad explorer and "digital" mathematics. In E. Faggiano & A. Montone (Eds.), *Proceedings of the 11th International Conference on Technology in Mathematics Learning* (pp. 149–155). Bari, Italy: University of Bari.

Lakoff, G., & Núñez, R. (2000). *Where mathematics comes from: How the embodied mind brings mathematics into being*. New York, NY: Basic Books.

Lyons, I., & Beilock, S. (2011). Numerical ordering ability mediates the relation between number-sense and arithmetic competence. *Cognition, 121*(2), 256–61.

Lyons, I., & Beilock, S. (2013). Ordinality and the nature of symbolic numbers. *Journal of Neuroscience, 33*(43), 17052–17061.

Lyons, I., Price, G., Vaessen, A., Blomert, L., & Ansari, D. (2014). Numerical predictors of arithmetic success in grades 1–6. *Developmental Science* [Epub ahead of print].

Noss, R., & Hoyles, C. (1996). *Windows on mathematical meanings: Learning cultures and computers*. Dordrecht, Netherlands: Kluwer Academic.

Papert, S. (1980). *Mindstorms: Children, computers and powerful ideas*. New York, NY: Basic Books.

Sinclair, N., & Heyd-Metzuyanim, E. (2014). Learning number with *TouchCounts*: The role of emotions and the body in mathematical communication. *Technology, Knowledge and Learning, 19*(1–2), 81–99.

Sinclair, N., & Jackiw, N. (2011). *TouchCounts* [software application for the iPad].

Sinclair, N., & Pimm, D. (2014). Number's subtle touch: Expanding finger gnosis in the era of multi-touch technologies. In C. Nicol, S. Oesterle, P. Liljedahl, & D. Allan (Eds.), *Proceedings of the 38th conference of the international group for the psychology of mathematics education and the 36th conference of the North American chapter of the psychology of mathematics education* (pp. 209–216). Vancouver, BC: Psychology of Mathematics Education.

Vergnaud, G. (2008). The theory of conceptual fields. *Human Development, 52*, 83–94.

PART IV

BROADER CONSIDERATIONS ABOUT DIGITAL CURRICULUM

CHAPTER 15

DIGITALLY ENHANCED LEARNING

Philip Daro

This chapter discusses how digital technologies can enhance learning from the perspective of what we have learned while designing, developing, and trialing a K–College and Career Ready mathematics curriculum for tablet platforms, Pearson System of Courses (PSoC). The basic premise of the program is that each student has the use of a tablet during class. The function of the program is to replace the textbook, not the teacher. Thus, this program differs from programs and apps where each student works alone at his or her own pace through a sequence of screens or immersed in a virtual world. Augmenting and enabling human feedback is, perhaps, the most important promise of digital technology. This program is designed for a classroom of students frequently interacting with each other and the teacher.

PRESENTATION OF CONTENT

In some long ago time, teaching and learning emerged as intentional activities, perhaps in pre-human species...modeling with feedback, no doubt. In human society, representational communication scaled up the complexity of what could be taught: language, art and diagrams, artifacts, dramatization and storytelling. Written language made it possible for teacher and learner to be asynchronous...cross generations and cultures. Institutions with a mission of teaching emerged: schools. Later, printing technology democratized reading and writing. As society depended more on educated members, schooling spread. Education became a means of social mobility for larger and larger portions of the population. And now, advances in technology are the largest since printing a half millennium ago. Access to and facilitated navigation of content has been transformed. We do not know the ultimate results of this impact. We can only see a few screens into a future that is still looming.

From Print Books to e-Books

Already, e-books are ubiquitous in higher education. The migration of e-books from higher education to secondary schools preparing students for college and career cannot be far away. As textbooks, e-books can provide layered and linked content and interactivity impossible with paper. Video clips with audio in addition to still pictures are just the beginning. Instead of static tables of data, the database itself can be aboard for re-analysis. Graphs can be generated to answer new questions. Simulations afford "readers" the experience of interacting with dynamic structures, not just interpreting static diagrams. And, not least, we have been liberated from the orthopedic limits on how much text one can carry around.

The boundary between reading and writing is moving. Reading has acquired new layers of reader agency, or at least reduced the effort needed. Annotation tools, cutting and pasting, and searches based on text excerpts are but a few reader acts that can cross over to writing acts. We need new expectations for accountability to the text, quotation, and "mixing" vs. plagiarism.

Print books afford spectacular ease of browsing and random access. E-books have some drawbacks. Scrolling is not as effective as flipping pages to browse; it is annoying. But print books do not have much in the way of *search* functions. The print index and table of contents are weak compared to e-book searching. Authors and readers are still learning how to exploit "search and find" features in e-books. Hyperlinks and navigation to other sources, local area and worldwide, explode into possibilities that dwarf printed citations and bibliographies. Readers of e-books can search deeper and wider into topics in real time. And, they can fetch content from elsewhere and bring it back.

Navigation and the Web

The easy navigation in real time to worldwide sources has already changed expectations for student assignments. Students can and are increasingly expected to find and cite evidence, alternative ideas, and data from sources far beyond what was available in print libraries and lost copies of journals. For the student, the relationship between reading and writing has become much more intimate. Sometimes too intimate: read and cut, write and paste.

Navigation to the web goes beyond a virtual library of text, e-text, and databases. A meta-world of standard courses is exploding. Mathematics has more than its share of standard courses with large enrollments also in service courses. These are markets for free and inexpensive learning assets easily found on the web via standardized taxonomies of topics and terms. Web videos of mini-lectures (MOOCs and Khan Academy are the tip of the iceberg) that provide an alternative to faculty lectures on the same topic. In K–12, there are videos of dozens of teachers and some students for many of the topics in school mathematics. These vary in mathematical and pedagogic quality, mostly toward the disappointing end of the scale. They are mostly well meant and good humored. Whether the quality improves may depend on whether any of today's evaluation and curation efforts gain traction and are effective.

The flood of web video lessons can lead to informal flipping: students becoming more dependent on explanations from online sources than on classroom explanations. A few teachers and schools have formalized flipping: get the lecture from video outside class, use class time to work on problems with teacher oversight and interaction. One can easily imagine good and bad versions of this scheme. In some ways, flipping is as old as the hills in other subjects: the study of literature and history has long (always?) been structured by heavy reading outside class and discussion in class. Perhaps the peculiar properties of the mathematics lecture and its relationship with the peculiar properties of mathematics homework lead us to think something new is afoot here.

Social Networking

Outside schooling, social networking has become a dominant use of technology and the source of large but poorly understood transformations of culture and society. It is prohibited in most schools. Can schools survive with this discontinuity between school and life elsewhere? As social networking gains larger shares of social reality, places without it lose reality and feel like social fictions imposed by the powerful on the powerless... not a motivating feeling. As access to content previously monopolized by schools becomes available through social networking and the web, the diminished social reality of time isolated from social networks will feel even less warranted. Whether this is good or bad, we do not know yet. We should probably be doing more by design while we wait to see.

Digitally supported learning communities are already possible, but not common for school subjects. When will students be able to call out worldwide for others... teachers, students, educated mentors... for collaborative learning on, say, quadratics? Anyone out there who can help me with problem 8, page 215? Learning communities can persist as members turn over. Communities of students learning linear functions can grow and shrink seasonally, and change members, as students progress through the curriculum. This happens outside school subjects, but so far, it is faint within school subjects for students. One example of a learning community for math teachers interested in interesting problems and learning from problems has been ignited by Dan Meyer via his blog, http://blog.mrmeyer.com.

Animation

The significance of *animation* for representation of complex ideas is this: time can be represented in time. That is, action in time can be portrayed directly. This has tremendous potential to enable visualization and comprehension of processes that unfold in time. Scientific phenomena, for example, typically unfold in time. In mathematics, structures that relate domains of values can be animated to show the relationships as values varying over time: functions and their lower grade precursors are obvious examples. As early as Kindergarten, calculations model actions on counted or measured quantities; pictorial and diagrammatic examples of these actions can be animated over time. The number line is fundamental to understanding number (especially fractions and decimals), operations, numerical measures of quantity, and ratios. Yet static diagrams can get complicated and confuse students in elementary grades. Reversible animated diagrams of the same actions on number lines can be so much more comprehensible. And of course, dynamic graphing software and computer algebra can be used to present content as well as do work for students.

PEDAGOGIC INTERACTIONS

What I Learned about Motivation from Game Developers

I had the chance to work with game developers while working on PSoC. These were young non-educators. Their job is to motivate kids to try the game and get hooked very fast. Hooked meant they didn't want to put it down, and, when they did stop, they were anxious to re-enter the game.

As the game developers talked about all this motivating the "player," I kept recalling they were talking about our students, the identical human beings. Don't we, educators, also worry about motivating them? Don't we also want them to want to re-enter mathematics? It struck me that we didn't talk about motivation in the same way. Our premise: mathematics is good for them and that's why *they have to learn it* and we have to make them. As the discussion moved forward, a dim shame was coming over the horizon.

I asked the gamers to explain how they motivated players. "The most important thing," they said, "is 'LTF,' (Low Threshold Feedback). That's the little noises, wiggles, and glows; the pops and tinkles, winks, and tumbles. Every time the player acts, the game responds, LTF; every second or so, tells the player your action causes the game action: you are the cause."

You are the cause. There it is. The motivation. My dim shame was glaring. I was remembering (from grad school) this: when Piaget was asked, "You describe how children progress in their intellectual development, but you never explain why, what motivates them?" He responded, "For the joy of being the cause." The joy of being the cause. *The joy of being the cause.*

Just think of the many ways we organize schools and classrooms to give each and every student the frequent feedback that they are the cause: your action causes the action here. Can technology help change the direction of causality of instruction and the frequency of feedback?

The second principle of motivation in games comes from the ways accomplishment registers. The badges and levels. Failure does not register. Never. Only success. It is a try, try again morality. It is a never quit morality. When a slow-to-learn player takes 14 tries to reach Level 3, he gets the same identical badge as his sister the whiz who got to Level 3 in one try. Like scouting, accomplishments count no matter how hard it was to accomplish. Most of the real world works that way, too. School doesn't.

The third principle is how the super-competitive players are engaged: top ten lists, highest score postings. These only apply to the top most players. Most of us are motivated by "your highest score." Games have three levels of feedback. For everyone, even the least player, the noises and wiggles of LTF. For most of us, the try, try again levels and badges. And for the few who are interested in competition at high levels, the top ten lists. The analogy in school would be very frequent feedback in response to very frequent student actions, badges for real but noncompetitive try-try accomplishments (up to A–), and a more competitive grading to distinguish A from A+.

Technology can enhance motivation by applying these principles beyond gaming. Beyond these principles, technology can broaden the palette of engagement on offer: social connections, the power to browse and find, to dig deeper at will, to specify mathematizable situations using dynamic visual and audio content well beyond the infamous text genre "word problems," and interactive simulations where the learner drives the action to accomplish self-set or program-set goals.

Cognitive Demand

One disappointing aspect in the many apps and programs available and aimed at K–12 is the low level of cognitive demand on the learner. Who is doing the mathematics? Students learn from their own actions, not from the actions of the program. Flat demand that doesn't go up and down becomes tedious, numbing cognition rather than stimulating it.

Students are learning to produce whatever the program assigns them to produce, under the conditions in which they work. If students spend most of their time producing answer selections from multiple choice questions, that is what they are learning. Selecting pre-written answers is far from writing answers into blank space. If students spend most of their time producing short answers, that is what they are learning. In mathematics textbook programs, an inspection of the "answers in the back of the book" shows at a glance what the students are being asked to produce. Good mathematics teachers have always asked students to produce more: justifications and arguments, explanations, alternative solution paths, co-ordinations across representations, and graphs from scratch ... in a word, mathematics.

When evaluating instructional material, no matter what the technology, ask first, "What are students assigned to produce?" Compare what they are assigned to produce with what your standards say they should learn to produce. We cannot assume the student's response is evidence of learning the mathematics in the stimulus. Being able to answer correctly the question, "How fast are we going?" by reading the instantaneous speed from a speedometer is no evidence of understanding Calculus.

Embedded assessments that have the same contextual properties as the pedagogic experience may not be assessing anything outside the scaffolding effects of those contextual properties; students might be only learning inside the teaching context with little or no transfer to other contexts. We need to be wary of claims about learning based on internal assessments.

Such assessments have value for telling students and teachers how well a student is progressing in the program, but this is not equivalent to progressing in mathematics learning.

Interactives and Games

A blank piece of paper and pencil is a wonderfully flexible interactive. The feedback in this familiar system is seeing what is written. Many of the representations we use in mathematics were developed and evolved for use in this paper and pencil system. Like any technology, it has its limits: its representations are static; many are laborious to produce, thus raising the cost of experimentation; copying is tedious; presentation tends to follow the chronology of production; and paper scatters easily in the student's world—dogs eat it.

The most serious limitation of paper and pencil is the range of feedback available. Aside from looking at what has been written, writing can be made public and thus support feedback from other people: peers, tutors, teachers. Human feedback is probably the most powerful and precious resource in the pedagogic economy. Written work enables a degree of asynchrony in the feedback systems, but the demand on respondent time is very large compared to the need. Augmenting and enabling human feedback is, perhaps, the most important promise of digital technology.

The power of scratch paper as a thinking aid cannot be overestimated and should not be abandoned when we pick up a tablet, but blended with new technologies. Indeed, one wonderful feature of tablets is the camera, which can take a picture of work on paper and commit it to navigable memory where the dog can't get it.

Interactives can set up immersive worlds where student action causes change in the world; student action has visible and audible consequences. The rules that govern the relationship between student chosen acts and the consequences can be designed to correspond to the rules of mathematics. The properties of the world can conform in designed ways to the properties of number and operations, the properties of equality and inequality, the properties of geometric transformations. Nothing is more natural in the video world than transformations.

The student acting and then evaluating the consequences of the action becomes the engine of learning in a game or interactive. The context for this engine can be unadorned mathematical representations, such as graphing and dynamic geometry tools and algebraic calculators. Or the context can be as richly skinned as a video game. And everything in between.

What Students Make: Discussion, Ideas, and Sense

New knowledge comes from engaging old knowledge in new situations where it doesn't make sense. Old answers become new questions; new questions motivate new answers... understanding new answers makes new knowledge. New ways of thinking are learned, in part, from making sense of the ways others think. This can happen if I am thoughtful and open-minded as I read Aristotle. It can happen more readily when I work to make sense of others in discussion... when I comprehend the different thinking of others. In discussion, I can also hear myself trying to explain how my thinking makes sense... comprehend my thinking as it appears to others. Even making sense of Aristotle or Euclid can be enabled by discussion with others. This is not the time to argue the importance of making sense and discussion for learning complex ideas, so we will take it as given.

Many initial dreams of digital learning ignored discussion and sense making in favor of individuals marching through screens or immersed in a game-world solo, "reading" content and answering questions which steered the student adaptively to new content. Students can zoom or slug through content at their own pace. Learning what? How is this different from different students reading the same book at their own paces, or even different books?

Regardless of the medium, students must be prompted by the instructional material to grapple with and make sense of complex problems and concepts. For most students, this grappling and making sense needs to be made concrete and visible. Their thinking made visible, not just the thinking of authors. For this to happen, most students need discussion. How can digital materials prompt, support, and enhance discussion? One way, if not the main way, is for the digital system to be designed for the students and teacher together: design for enhancing, not diminishing, the teacher's role in leading discussions and bringing the mathematics into view.

What Students Make: Assigned Work

Students learn to make what they make. What they are assigned to produce is a critical marker of what they are learning. The Common Core State Standards (National Governors Association Center for Best Practices & Council of Chief State School Officers, 2010) and other advanced syllabi ask students to make more than answers; they are expected to make and critique viable arguments and explanations that make sense to others. What is the impact of digital technologies on what students are assigned to make?

The Common Core State Standards and similar world-class standards expect students to produce well-reasoned arguments, explanations of why an approach or solution makes sense, justifications for assumptions in mathematical models, models themselves, and mathematical representations. They expect students to define functions and more that goes beyond mere "answers" or solutions to problems. What is the role of technology in the production of this assigned work? Here is where understanding digital tools as *productivity* tools is important. When we see the students as agents in the pedagogic process, the productivity functions become valuable assets.

In PSoC, students are routinely assigned to prepare presentations to the class that explain and represent their mathematical thinking...a regular episode in the lesson design called "ways of thinking." The teacher selects three or four of these prepared presentations for whole class discussion. The tablets are basic tools for creating presentations. Apps and embedded tools can generate graphs and diagrams, enable and enhance collaboration and feedback, provide tools for search and reference, ease revision and commentary; the camera can capture work on paper and even short video clips produced by students. The student as maker is a student operating with self-regulation and motivation that alters the dynamic of classroom culture when compared to activities where the students are passive or objects of assessment.

With the student as maker, the pedagogic process resembles writers' workshop. Writers' workshop is a method for organizing classrooms to do the work of writing. The writing process involves stages of prewriting (thinking and talking about a topic), drafting, getting responses to drafts from readers, revision, editing for correctness, and publishing. A writers' workshop organizes the classroom into definite locations where each of these writing processes take place. As a student progresses through the writing process, they move to the appropriate location where they join others at the same stage who are available for collaboration. There is a rhythm between working alone and collaboration that is driven by progress with the task at hand. These classroom routines are familiar to any student who has had a teacher who uses writers' workshop methods to teach writing.

When similar routines are used in mathematics, as with writing, it is the work that is analyzed and evaluated, not the student. Technology is not necessary to organize mathematics instruction around the student as maker...indeed, instruction with student as maker is done on a large scale in Japan and in other Asian systems in low-tech classrooms. But technology in the hands of a student as maker is technology used the way it is used everywhere outside school. This use contrasts sharply with technology as content presenter interleaved with assessment, in which the student is the object of instruction and assessment rather than the agent.

Motivation research by Carol Dweck (Dweck, 2006; Mangels et al., 2006),[1] Andrew Elliot (Elliot & McGregor, 2001), Catherine Good (Good, 2006), and others have established the benefits of establishing the students' beliefs in themselves as agents of their own learning as a means of becoming "better at math." This research also shows the dangers and costs of feeding the belief that the student has a fixed trait of ability, which is a serious challenge to assessment models based on measuring stable traits rather than changing abilities through learning. We should design instruction to enhance the sense of agency that comes with students using technology to make things instead of merely using technology to make measurements of students, diminishing their sense of agency in becoming better at math.

The Writing Process: A Pedagogic Model

The *writing process* was developed in the early 1970s by the Bay Area Writing Project. Since then, the Writing Project model of professional development and the writing process and writers' workshop models for organizing instruction in writing in the classroom have spread to hundreds of university based Writing Projects and over 2 million teachers. This is an important example of "going to scale" that should not be overlooked. For our purposes, it demonstrates the scalability across a wide variety of teachers of the pedagogic model for teaching writing. The properties of this model were core design principles for PSoC. These properties are discussed briefly here as they emerge in the model for using technology to teach mathematics.

It is no coincidence that basic lesson design for PSoC is an adaptation of Japanese lesson design to the American teacher experience with writers' workshop. The Japanese studied the writing process and workshop model in the United States to inform their own designs for teaching writing. They adopted and adapted. Many pedagogic processes have been migrating from writing to other curriculum areas, especially in the context of teaching disciplinary writing. This was true for writing in mathematics.

In Japan...indeed, most Asian countries...students were traditionally called upon to make formal, prepared presentations to the class. These presentations were a genre of writing. The genre in mathematics was essentially a mix of explanation and argument. As part of this tradition, critique and questioning from other students and the teacher were typical. The writers' workshop processes grafted naturally to this tradition with surprising and wonderful results. Table 15.1 shows how it works from the perspective of mathematics "writing": student presentations, in contrast to traditions of mathematics teaching in the United States.

In developing their presentations, Japanese students routinely present drafts to partner students and make revisions based on the response of their partners. The standards for what students produce and what they make are much higher in the Japanese design, as they are in American writing process instruction. From this perspective, the American tradition in mathematics instruction usually settles for first draft, improvised explanations. If

TABLE 15.1 Student Presentations in U.S. Tradition Compared to Japanese Design

U.S. Tradition	Japanese Design
The audience has been the teacher in the role of checking the student.	The audience is the other students in the class, so they will understand the presenter's "way of thinking."
The purpose has been to show "I am a good student, paying attention."	The purpose is to produce an explanation/argument that makes sense to the other students.
The students are volunteers using the "ooh, oooh, ooh … " stabbing hand-raising motion to volunteer as fast as possible (and hands on the desk method of avoiding participation).	Everybody ready: all students prepare, in writing, a presentation. The teacher selects presentations based on inspecting the writing. The selections (3 or 4) represent the different ways of thinking evident across the class.
The presentation is improvised and brief, sometimes less than a sentence … rarely a developed argument.	The presentations are complete arguments and explanations, with visual rendering on the board of mathematical representations and diagrams.
The extent to which other students understand the presentation is given little attention.	The teacher checks how well the other students understand the way of thinking. The teacher pushes for revisions: increased precision of language and explicitness of reference, illustrations and examples "so more students will understand."
The flow of ideas as represented by these improvised fragments is haphazard even as teachers heroically try to weave the fragments into coherent cloth.	The flow of ideas is organized by the teacher by sequencing the presentations and focusing on correspondences across ways of thinking. The teacher summarizes the mathematics explicitly, citing the written text of the student explanations on the board.
No process and no time are allocated for students to revise their explanations or correct them.	The teacher summarizes and edits for correctness anything that hasn't been corrected during revision.

we think of the explanations as manifest understanding of the mathematics, students are not expected to revise and edit their understanding as a routine part of the pedagogic process. Whether they do or not is only evidenced by correctness of answers to problems.

In PSoC, we adopted the Japanese design, with the extensive guidance of Akihiko Takahashi and Tad Watanabe who served as lead designers of the K–5 courses. We framed this lesson design in familiar writers' workshop structures. Taking this basic design as a given, we asked, "How can digital technology enhance this pedagogy?" We have already discussed how using interactivity and video capability enhanced presentation of the problems (i.e., the student prompts). We also discussed how the productivity functions of tablets empower students as makers of their own explanations/arguments.

This pedagogy of writers' workshop synthesized with Japanese lesson design leverages the variety of ways students think as learning resources. Student to student interactions, teacher guided discussion, and summary can progress through the students' ways of thinking from most concrete to more abstract and be aligned with grade level goals. The technology enhances these interactions in ways familiar from everyday life.

- Work can be posted and shared at every stage of development.
- Students use annotation features to comment on each other's work.
- Revision is convenient.
- The camera brings in production off the tablet.
- Memory and browsing generate a corpus of work that serves as a resource for reference and study.
- Video enables dynamic representations and captures otherwise ephemeral discussions.
- Workflow management tools help students keep track of their work.

CONCLUSION

Digital technology invades our lives because we welcome the extension of our own agency...it makes us more powerful. We can do and make things previously beyond our reach. It serves our purposes at such low costs that we enjoy making trivial and momentary delights. Fixing mistakes is often so easy that spontaneity and experiment flourish. We use it to create and present extensions of our identity. This technology should invade the classroom for these same reasons.

The technology is also a labor-saving appliance. It has enormous capacity for memory. We can make things, find and clip things, collect, cut and paste, mix and match to build a corpus to be used in our work. We can search this corpus to study, solve problems or make new things. We can share and comment upon our own or others' collections. We can receive or make assignments, keep track of our work, manage assignments, and carry it with us.

A major issue for technology in schools is the culture clash between educational institutions and the digital world. For whatever reason, the culture of educational institutions has long been committed to stodgy-style in its artifacts and communications. Stodgy-style lacks self-irony and is oblivious to its own pomposity. Its taste in "positive" expression is kitschy, selectively nostalgic for "tradition." Stodgy-style is characteristic of authoritarian organizations. The culture of digital technology is quite different: adolescent humor is ascendant; self-irony, irreverence and iconoclasm are pervasive. Inventiveness and freedom trump taste and convention.

An example from the PSoC experience. When we looked at story problems written by middle school boys, they were like movie trailers or video games. If train A was overtaking train B, then train B had bad guys and a bomb in it, while train A had the hero with superpowers. Constraints on the superpowers figured into the quantitative structure of the problem. There were fights and explosions. None of this motivating, entertaining (for teen boys anyway) context conforms to social norms for educational materials in schools. We also saw shopping problems that looked nothing like our traditional word problems. The mall setting and current fads and fashion were dominant themes. The characters' motives had more to do with achieving goals related to relationships (boyfriends and girlfriends) than they did with how many people were sharing cookies at auntie's house. In PSoC, we were able to capture some of this spirited, often comedic culture, particularly in the video clips. It helped to go to outside education providers. But we were feeling the limits of "tradition" around every turn. The lack of self-irony in "political correctness" was an especially wet blanket.

How can we bring the humor and spirit of the adolescent world into schools that are charged with bringing adolescents into the adult world? We need a better answer than the dull correctness now dominant. Giving students more agency in the pedagogic scheme and allowing them to import the humor into classroom life is a good bet. Digitally-enhanced agency through the productive powers of technology is exciting: student-made videos, for example. It must be noted here that "high performing" countries are also stuck in a stodgy world. It wouldn't surprise me to see the United States lead the way here. Toshiakira Fujii (personal communication) said the Japanese society is more closed than the U.S. society, but the Japanese classroom is more *open* than its surrounding society. He asked, "Why is the American classroom so much more closed than its surrounding society? Isn't it difficult for students to stay engaged?"

The irreverent culture of the digitally-enhanced world in which students live will migrate into our classrooms with or without educator participation. We can capitalize on the enhanced productivity of workers using new technologies only to the extent we enable the agency of students and teachers as learning workers. Perhaps the arrival of personal technologies will be the stimulus for a long overdue freshening up of our pedagogic and management systems so the humanity of students becomes a source of energy and strength for learning instead of a nuisance.

NOTE

1. See also the Brainology website at http://www.mindsetworks.com/webnav/program.aspx.

REFERENCES

Dweck, C. S. (2006). *Mindset: The new psychology of success.* New York, NY: Random House.

Elliot, A. J., & McGregor, H. A. (2001). A 2x2 achievement goal framework. *Journal of Personality and Social Psychology, 80*(3), 501–507.

Good, C. (2006). Transforming classroom culture through the use of student allies. *Network News: Newsletter of the Minority Student Achievement Network, 22.* Evanston, IL: MSAN.

Mangels, J. A., Butterfield, B., Lamb, J., Good, C., & Dweck, C. S. (2006). Why do beliefs about intelligence influence learning success? A social cognitive neuroscience model. *Social, Cognitive, and Affective Neuroscience, 1,* 75–86.

National Governors Association Center for Best Practices & Council of Chief State School Officers. (2010). *Common Core State Standards for Mathematics.* Washington, DC: Authors.

CHAPTER 16

MATHEMATICS STANDARDS AND CURRICULA UNDER THE INFLUENCE OF DIGITAL AFFORDANCES

Different Notions, Meanings, and Roles in Different Parts of the World

Mogens Niss

INTRODUCTION

In the field of education, and in the field of mathematics education in particular, the world is much more diverse and complex than we often tend to think. Many things that are taken for granted in one context turn out to be surprisingly different in other contexts. This also pertains to the content of established terms, such as *standards*, *curriculum*, *guidelines*, and *syllabus*, to name just a few. Not only do these terms have rather different meanings in different countries, the entities referred to by the terms play very different parts as well.

What can we learn from such, at first confusing, diversity and complexity? First, we note that things that vary across contexts and conditions may, in principle, be changed, whereas things that remain constant are likely to be harder to change. Knowledge of and insight into what is *variable* and what is *constant* contributes to making us sharper and wiser. Second, understanding the ways in which our own situation differs from that of others helps shed light on the specifics of our situation, and, hence, on the potential for changes to it, should change be deemed desirable. Third, if there is a need for change in our own situation, we can get inspiration for our endeavor by studying other approaches and solutions and the conditions under which these have been undertaken. These points underpin the exposition that follows.

THE NOTION OF CURRICULUM

Let us begin by taking a closer look at the very notion of *curriculum*, which seems to be used in most countries, even though translations into national languages carry lots of different, and not necessarily mutually compatible, connotations. So, not only does the very notion of curriculum vary greatly across countries, so does the "reality of the curriculum" (i.e., both the reality within which a given curriculum lives and the reality created by a given curriculum).

If we consult a standard English dictionary, such as *Collins Cobuild* (1999), we may find the following general definition, "A curriculum is all the different courses of study that are taught in a school, college, or university" (p. 401). In contrast to this institutionally oriented definition, focusing on the entire collection of the courses of study in an institution, Kilpatrick (1994) offers a definition which focuses on the substantive meaning of the term *the curriculum*: "The curriculum can be seen as an amalgam of goals, content, instruction, assessment, and materials" (p. 7).

Alternatively, Stein, Remillard, and Smith (2007) provide a definition of the mathematics curriculum as primarily seen from the teacher's perspective: "...we use the term curriculum broadly to include mathematics curriculum materials and textbooks, curriculum goals as intended by the teachers, and the curriculum that is enacted in the classroom" (p. 319, footnote).

In order to establish a platform for subsequent analysis and discussion, I find it worthwhile to propose a general definition, albeit a somewhat complex one, of the notion of curriculum, pretty much in line with that of Kilpatrick. By *an educational setting* I understand the structural, organizational, and temporal unit within which a certain package of teaching and learning of some subject unfolds. Thus, an educational setting may range from consisting of all compulsory public schools in a country in a given period of time, over being a single educational institution, say a school or a university, to consisting of a specific stream amongst all streams in, say, an upper secondary school in the spring term of 2015, or of a particular mathematics course offered by a university department in the current semester.

Even though the following discussion might, to some extent, be generalizable to any educational subject, I concentrate on mathematics education and mathematics curricula. In my understanding, a curriculum refers to a given educational setting, whether extensive or limited. I propose to define a *curriculum* with respect to such a setting as a *vector* with six entries:

- *Goals* (the overarching purposes, desirable learning outcomes, and specific objectives and aims of the teaching and learning taking place under the auspices of this curriculum);
- *Content* (the topic areas, concepts, theories, results, methods, techniques, and procedures dealt with in teaching and learning);
- *Materials* (the instructional materials and resources, including textbooks, artifacts, manipulatives, and IT-systems employed in teaching and learning);
- *Forms of teaching* (the tasks, activities, and modes of operation of the teacher of this curriculum);
- *Student activities* (the activities of and tasks and assignments for the students taught according to this curriculum); and
- *Assessment* (the goals, modes, formats and instruments adopted for formative and for summative assessment in this curriculum).

Introducing or defining a given curriculum within an educational setting then means specifying the entries represented by these six components. Analyzing an existing curriculum amounts to analyzing these six entries, as explicitly or implicitly present in the curriculum. I call the agency that defines a certain curriculum and has the actual or potential power to implement it within some jurisdiction, the *curriculum authority* for that curriculum. A curriculum authority may choose to leave some of the six entries empty when defining a curriculum. As will be exemplified below, curriculum authorities operating with respect to a given educational setting oftentimes form a partially ordered hierarchy.

The term *standard* (or "curriculum standard") does not seem to be a term that is universally applied. Prime examples of countries that have adopted this term are the United States and Germany (the "Bildungsstandards" [see, e.g., for the high school level, Kultusministerkonferenz, 2012], which can be translated into "educational standards"). On closer inspection, "standard" seems to have two different meanings. First, in some cases "a set of standards" means a *mandatory curriculum* for a given educational setting (say, primary mathematics education in some country or state), defined by a primary (i.e., overriding) curriculum authority at the top of the hierarchy of curriculum authorities for this setting. Typically, standards in this sense put emphasis on, as a minimum, the vector entries of goals, content, and (summative) assessment. The other prevalent meaning of "a set of standards" is a *recommended curriculum*. A recommended curriculum can be defined by any type of agency, for instance an individual or an association or a council, as was the case with the two National Council of Teachers of Mathematics (1989, 2000) standards and with the Common Core State Standards (National Governors Association Center for Best Practices & Council of Chief State School Officers, 2010) initiative in the United States. If a curriculum authority decides to adopt a set of standards in this sense it becomes mandatory for the jurisdiction of the authority.

CURRICULUM AUTHORITIES IN DENMARK

For a given educational setting, two key questions deserve attention:

- What are the different curriculum authorities functioning within that setting?
- What are the relationships amongst these authorities and amongst their respective curricula?

Examples of answers to these questions can be found in the case of Denmark. Similar answers are valid for Finland, Norway, and Sweden, although there are, of course, several differences when it comes to details.

For schools in Denmark, the primary curriculum authority is the government, implementing the decisions of the parliament. The government is represented by the Ministry of Education, acting on the government's behalf. In contrast, a university is basically its own primary curriculum authority, even though the government has the final decision on which study programs are accepted in each university, reflecting the fact that all Danish universities are public. For a given educational setting, say the high schools in the country, the government's official curriculum is mandatory and specified in written documents. There are four different high school branches, each of which offers two or three levels of mathematics. There is a *ministerial curriculum* for each combination of branch and level. Such a curriculum has the following shape:

- *Goals*: The goals are manifest and explicit. They are of an overarching nature and are stated somewhat tersely. For example, it is a goal for one of the branches that students should be able to "explain mathematical reasoning and the deductive aspects of mathematical theory."

- *Content*: The specification of content for each branch × level is very manifest and pretty substantial, yet kept at a general level. There is a strong emphasis on the role of CAS (computer algebra systems) and graphing calculators in the definition of the content.
- *Materials*: This entry is basically empty, except for the requirement that students are expected to buy and use CAS and graphing calculators.
- *Forms of teaching*: This entry is almost empty, except for a few guidelines concerning teachers' obligations to assign and correct written tasks and to give marks at regular intervals.
- *Student activities*: This entry is almost empty, except that students have to be assigned certain kinds of tasks at various stages through the high school years.
- *Assessment*: The specification in this entry is pretty substantial when it comes to summative assessment. All branches have end-of-program national written exams set by the Ministry of Education. The exams never contain multiple choice questions, and students' results are graded by a large corps of experienced teachers, who give out high-stakes marks. In the written exam questions, there is an emphasis on the use of CAS and graphing calculators. On top of that, some high school branches have final oral exams involving both internal and external examiners. The oral exams are organized by the ministry, which sends the external examiners to the schools and also sets guidelines for the procedures around and within the oral sessions. The final national exams are high-stakes exams, as the marks gained on them strongly determine the tertiary programs to which students with a high school diploma are admitted. There are no guidelines for formative assessment in the ministerial curriculum.

We may summarize this ministerial vector in symbolic form:

$$(\text{Goals; Content; Materials; Forms of Teaching;}\\ \text{Student Activities; Assessment}) = (m; M; 0; \varepsilon; \varepsilon; M),$$

where m stands for "manifest," M for "very manifest," ε "for very little," and 0 for "empty."

In Denmark, the ministerial curricula are mandatory—Americans may want to call them "national standards"—but as some of the entries are sparsely defined or empty, the ministerial curriculum leaves a fair degree of leeway and freedom to the teacher, which the teacher is left to utilize. This means that within the boundary conditions of the mandatory ministerial curriculum, the individual teacher has to define his/her own curriculum, which is typically enacted rather than written. The *teacher's curriculum* focuses on filling in those aspects of the entries that have been left empty or sparsely defined by the ministerial curriculum. Thus, the teacher assumes the role of an indispensable curriculum authority, albeit subordinate to the primary curriculum authority, the Ministry of Education. The entries of the high school mathematics teacher's curriculum then look like this:

- *Goals*: The teacher's curriculum has to respect the overarching goals in the ministerial curriculum. The teacher will usually add goals of his or her own, but typically these will remain implicit.
- *Content*: The teacher's freedom is limited to specifying the content in the ministerial curriculum formulated in general terms and to make amendments to it if he or she so wishes, and when there is room for it.
- *Materials*: This is a substantial component of the teacher's curriculum, as the teacher (most likely in collaboration with colleagues) has to select or devise textbooks, notes, worksheets with tasks, artifacts, and manipulatives, whilst also making use of ICT (information and communication technology) according to the ministerial curriculum.
- *Forms of Teaching*: This entry is left almost completely in the hands of the teacher, modulo a few boundary conditions set by the corresponding entry in the ministerial curriculum. For example, as part of his or her teaching, the teacher must, from time to time, design and orchestrate group project work or tasks dealing with overarching themes.
- *Student Activities*: This entry also is left almost completely in the hands of the teacher, modulo a few boundary conditions set by the corresponding entry in the ministerial curriculum.
- *Assessment*: Formative assessment is left completely in the hands of the teacher. By and large the same is the case with summative assessment, within the boundaries of the ministerial curriculum. As to the final national exams, as previously outlined, the teacher is not involved in the written part but is the director of the oral part, in those cases when it takes place, in collaboration with the external examiner chosen by the ministry. Using the same symbols as before, we can symbolically condense the teacher's curriculum as follows:

(Goals; Content; Materials; Forms of Teaching; Student Activities; Assessment) = (ε; m; M; M; M; M)

Within the same educational setting, and even within the same school, different teachers tend to define very different teachers' curricula. The same teacher may well define several different curricula for his or her different school classes, even within the same branch of mathematics and at the same level. There is no hierarchical relationship between these curricula, but they are all subordinate to the ministerial curriculum for that setting. To summarize the situation for high school mathematics in Denmark, there are two categories of curriculum authorities: the primary authority—the Ministry of Education—and the category consisting of individual teachers, each of whom is his or her own curriculum authority.

At the primary and lower secondary levels, the situation is slightly more complicated given there are three categories of curriculum authorities. In addition to the Ministry of Education and the individual teacher, the municipality (the local government, which is financially in charge of primary and lower secondary schools) is a curriculum authority as well, situated between the individual teacher and the ministry. However, in practice a municipality only seldom adds further specifications and requirements to the ministerial curriculum. As it does control the resources made available to schools and the number of lessons allotted to each subject on top of the ministerial minimum, the municipality heavily influences the boundary conditions for the individual teachers' curriculum.

In Denmark, there are no private curriculum authorities. Of course, one might perceive a textbook (or a textbook system) as specifying a rudimentary curriculum, which is massively focused on "content," often supplemented with a number of proposed "student activities" and "assessment" tasks as well. However, the other components, such as "goals" (other), "materials," "forms of teaching," "student activities," and general "assessment," are largely absent. This reflects the fact the textbooks are commercially produced and sold and need to be accepted and chosen by institutions and teachers so as to fit into their curricula. It is interesting to contrast this role of the textbook with that often found in the United States, where comprehensive paper-based or electronic textbook systems accompanied by specific online resources, often of an interactive nature, recommendations and materials for teachers, proposed student activities and assessment tasks, and sometimes in-service programs for teachers (e.g., summer institutes) are part of the package as well. This means that such textbook systems constitute full-fledged curricula in and of themselves.

In several, if not most, European countries, the national or provincial government (via the relevant Ministry of Education) is the primary curriculum authority for public schools, to which other curriculum authorities, including teachers, are subordinate. However, in some countries private/commercial curriculum authorities abound within independent private schools. Without claiming any extensive expertise, I believe the situation is not much different in many East Asian countries, such as Japan, Singapore, and South Korea. Within this general scheme, the room for and the role of the teacher's curriculum varies considerably across countries, ranging from relatively little to extensive curricular freedom.

DILEMMAS AND BALANCES

The specification of each of the six entries that define a curriculum can take an infinitude of different forms. It is clearly not possible in this chapter to go into any detail with them. Instead, let me take a closer look at some balances to be struck and some dilemmas encountered in the context of a few of the entries.

As to *goals*, they were, in the past, often implicit, or at best very terse and general. Today, different curricula display (at least) four types of goals. One predominant type is building "static" subject matter knowledge with students. Another type of goal—though less predominant than in the past—is to develop certain sets of procedural skills with students, whilst a third type focuses on developing "dynamic" mathematical competencies (or capabilities or proficiencies, if you wish) with students. The fourth type of goal puts emphasis on fostering students' attitudes, emotions, and beliefs regarding mathematics. These types of goals are very different in nature, scope, and emphasis without in any way being incompatible. A pertinent issue, on which one can find much disagreement, is whether the goals involve aspects beyond "pure" mathematics, such as learners' ability to put mathematics to use in extra-mathematical contexts and situations. The same is true of the issue of whether it is an independent goal for mathematics education to be a vehicle for general ICT familiarity and proficiency, and whether it is a goal for mathematics education to be a vehicle for developing general intellectual or personal traits such as logical thinking or perseverance.

When it comes to *content*, classically, lists of content elements have constituted the core of mathematics curricula. There are different categories of content, in addition to internal mathematical subject matter (topics, concepts, results, theories). These include mathematics as it is manifested within other subjects or practice areas; mathematical processes perceived as objects (e.g., proof); historical, philosophical, or sociological aspects of mathematics; and ICT systems and tools, amongst others. Again, these content categories are certainly compatible, but they are very different. The presence or absence of any one of them in the "content" entry of a curriculum is a significant characteristic of that curriculum.

In the past, in most places the *materials* entry comprised, above all, textbooks and task booklets, but also tables and sometimes physical tools such as rulers, compasses, protractors, slide-rules, and abacuses; in some cases also geometric models made of wood or plaster, and counting blocks. Today, curricula include several other kinds of materials: a wide spectrum of paper-based or electronic texts; a wide array of manipulatives, physical instruments (e.g., robots), and artifacts, including games; and long lists of hard and soft ICT systems, including computers, CAS and graphing calculators, tablets, cell phones, the Internet, and interactive media linking several sorts of materials. Generally, the different kinds of materials tend to become increasingly integrated. It is a characteristic feature of this development that the materials entry tends to evolve independently of the other entries as a consequence of external factors governing the general development of media, which is much faster than the development of curricular thinking and practices, which are then doomed to perpetually lag behind the development of materials. The fact that today more and more textbooks are available online in formats that allow users to pick sections from different books and combine or change them raises the issue of the balance between, on the one hand, expositional logic and coherence and, on the other hand, flexibility obtained at the cost of disconnected patchwork presentations.

Finally in this section, I briefly consider the sixth entry in the curriculum vector, *assessment*. Classically, in many places, yet not everywhere, "assessment" was equivalent to "testing." Today, we are faced with a multi-faceted array of assessment modes, instruments, and formats that have been put into practice to provide for the need to assess, in formative as well as in summative assessment, a wide spectrum of mathematical competencies in a multitude of different contexts. Researchers of assessment in mathematics education have realized that no single assessment mode, instrument, or format is adequate for assessing all significant components of mathematical learning, so there is no "one size fits all." Moreover, since there is ample evidence that "what you assess is what you get" (wyawyg), the "assessment" component of the curriculum is a marked determinant of the outcome of mathematics learning, and hence, has to be aligned with the "goals" component in order to prevent a mismatch from arising.

Concluding this section by a look to the place of digital technologies in mathematics curricula, it follows from what has been said that digital technologies may have an actual or potential role to play with respect to all of the six curriculum entries considered. The fundamental question is "What role?"

THE ROLE AND IMPACT OF DIGITAL AFFORDANCES

The role and impact of digital technology, in most of its facets, have been subjected to a huge amount of research and development during the last several decades. One outcome of this is the finding that the very same piece of digital technology can give rise to *"marvels"* as well as to *"disasters"* in mathematics education. This means that no ICT system, hard or soft, is, in and of itself, good or bad for mathematics education. The outcome crucially depends on the role and place of technology in the entire curriculum, including each of the six components previously considered, and on the specific relationships that exist between that component and other curriculum components, as well as on the teacher's design and implementation of the teaching-learning environment and of the instructional sequences that (are supposed to) take place within this environment. Of course, the outcome also depends on the nature of the digital affordances offered by the ICT systems at issue and on their technical and pedagogico-didactic quality.

Looking at the ways in which digital technologies are employed in mathematics education, two distinct purposes for their involvement become manifest, purposes that are closely related to the "goals" component of the curriculum. Digital technologies may serve to:

- enhance a wide variety of mathematical capacities, and
- replace some mathematical competencies.

In what ways can ICT enhance students' (and others') mathematical capacities? Amongst several other things, ICT can:

- help generate student experiences of mathematics-laden processes and phenomena that might be difficult to obtain by other means;
- create platforms and spaces for exploration in which mathematical entities can be investigated through manipulation and variation;
- produce static and dynamic images of objects, phenomena, and processes that are otherwise difficult to capture and grasp;
- create connections between different representations of a given mathematical entity;
- help solve hard or otherwise inaccessible computational problems;
- perform rule-based symbolic transformations and manipulations;
- support the production of mathematical texts; and
- create platforms for individualized training and assessment.

For most of these capacities to come to fruition in mathematics education, it is essential that the user understands the mathematical fundamentals of what is going on. By this I do not refer to the structure and programming of the underlying ICT systems but to the mathematical entities that are subject to treatment and display.

But there are also things that ICT cannot do for mathematics education. ICT cannot:

- replace students' creation of meaning and understanding of mathematical concepts and results;

- replace reasoning and sound and critical judgment;
- replace problem-solving competency;
- replace symbols and formalism competency, including the ability to perform basic computations;
- construct, interpret, or validate mathematical models; and
- replace the work needed to understand "what?," "how?," and "why?" in mathematics.

Digital affordances contribute to creating marvels in mathematics education, when ICT is part of didactico-pedagogically thoughtfully designed teaching-learning sequences in mathematics, in which the specific purpose and role of ICT in capacity enhancement is clear and articulate, in which the division of labor between ICT and other components is explicit and well-founded, and in which the teacher is fully aware of what ICT can—and cannot—be expected to contribute to the context at issue.

Digital affordances contribute to creating disasters in mathematics education when the purpose and role of ICT and its relationships with other components of a teaching-learning sequence are not clearly and carefully thought through and accordingly implemented, when what is happening is haphazard, when the ICT system employed is allowed to carry students and teachers away beyond its purpose for mathematics teaching and learning in the given context (for instance, when coming to grips with the system takes an excessive amount of time), in short, when mathematics education "cannot afford digital affordances." Disasters tend to occur when mathematics education comes too much under the influence of *"itealists"* (yes, the "t" is intended), who end up furthering ICT for its own sake rather than furthering mathematics education by way of ICT, in other words when the tail is allowed to wag the dog.

REFERENCES

Collins Cobuild English Dictionary. (1999). London, England: Harper Collins.

Kilpatrick, J. (1994). Introduction to Section 1. In A. J. Bishop, K. Clements, C. Keitel, J. Kilpatrick, & C. Laborde (Eds.), *International handbook of mathematics education* (pp. 7–9). Dordrecht, Netherlands: Kluwer Academic.

Kultusministerkonferenz. (2012). *Bildungsstandards im Fach Mathematik für die Allgemeine Hochschulreife* (Beschluss der Kultusministerkonferenz vom 18.10.2012). Retrieved from www.kmk.org/fileadm/veroeffentlichungen_beschluesse/2012(2012_10_18-Bildungsstandards-Mathe-Abi.pdf

National Council of Teachers of Mathematics. (1989). *Curriculum and evaluation standards for school mathematics.* Reston, VA: Author.

National Council of Teachers of Mathematics. (2000). *Principles and standards for school mathematics.* Reston, VA: Author.

National Governors Association Center for Best Practices & Council of Chief State School Officers. (2010). *Common Core State Standards for Mathematics*. Washington, DC: Authors.

Stein, M. K., Remillard, J., & Smith, M. S. (2007). How curriculum influences student learning. In F. K. Lester (Jr.) (Ed.), *Second handbook of research on mathematics teaching and learning* (pp. 319–369). Charlotte, NC: Information Age.

CHAPTER 17

MATHEMATICS CURRICULUM, ASSESSMENT, AND TEACHING FOR LIVING IN THE DIGITAL WORLD

Computational Tools in High Stakes Assessment

Kaye Stacey

INTRODUCTION

Digital technology provides mathematics education systems with a computational infrastructure and a communications infrastructure (Stacey & Wiliam, 2013). For the last decade, many teachers of all subjects in Australia have embraced the new opportunities arising from digital technology as communications infrastructure. These opportunities include: presenting students with lesson materials and assessment items online; using electronic textbooks which often include animations and linked multiple representations; showing short demonstrations and explanations of mathematical ideas; having students conduct Internet research or research in remote collaboration with external experts (e.g., a scientist); and enhancing communication from student to teacher, between students within the classroom or beyond it, and from students and teachers to parents. These are exciting developments, which have received strong endorsement by education authorities and governments (Government of Australia, 2013) and are gradually changing everyday practice in our schools. As is evidenced by the program for this conference, similar innovations are happening in many countries around the world.

This chapter, however, focuses on progress made by using digital technology as a computational infrastructure for curriculum, assessment, and teaching of mathematics. The opportunities and challenges of using digital technology as communications infrastructure are fundamentally similar for teachers of all subjects, including mathematics. However, because computation and mathematics have a unique relationship, using digital technology as a computational infrastructure presents opportunities and challenges specific to mathematics teachers, potentially changing the nature of the subject in fundamental ways. The subject is changed both because old mathematics has been changed by new computational options, and because the growth in computational power has enabled mathematics to be applied to completely new problems and has stimulated the development of new mathematics. When digital technology was first accessible to schools (e.g., in the 1970s and 1980s), mathematics was the subject most affected. Now, the energy and major impetus for change due to digital technology is across the curriculum, through the communications capabilities, but the special questions for mathematics still remain incompletely answered. Furthermore, the answers that we do have are incompletely implemented.

The chapter begins with a survey of the use of advanced computational tools in senior secondary mathematics in my home educational system, demonstrating how tools have been successfully incorporated into the curriculum, assessment, and teaching and are now accepted as normal. It reports extensive research that shows a small but consistent improvement without disadvantage. It is, however, the case that their adoption has led to this positive improvement without the radical rethinking that is required to continue to keep senior school mathematics relevant for students' futures. I report on a series of studies of student performance with different technologies that provide unique insight into some of the issues of technology change and end with some reflections on achievements and challenges to address.

CHANGING TECHNOLOGY FOR MATHEMATICS IN VICTORIA

Table 17.1 is adapted from Leigh-Lancaster (2010). It shows the evolution of the technology permitted to be used by students in the state examinations

TABLE 17.1 Assumed and Permitted Technology for Year 12 Examinations (Victoria, Australia)

Stage	Assumed Digital Technology	Notes
Further Mathematics ("Elementary" Year 12 Subject)		
–1996	Scientific calculators	Subject began around 1990
1997–1999	Scientific calculator with bivariate statistics functionality	Graphics calculators permitted
2000–	Graphics calculators	CAS calculators permitted from 2006
Mathematical Methods ("Intermediate" Year 12 Subject)		
–1978	Logarithm tables, slide rules	
1978–1996	Scientific calculators	
1997	Scientific calculators	Graphics calculators permitted
1998–2005	Graphics calculators	CAS assumed in MMCAS pilot from its inception in 2002
2006–2009	Examination 1: No technology Examination 2 (MM): Graphics calculators	Examination 2 (MMCAS): CAS calculators or software assumed
2010–	Examination 1: No technology Examination 2: CAS calculator or software	New subjects aligning with Australian Curriculum from 2016 with unchanged technology.
Specialist Mathematics ("Advanced" Year 12 Subject)		
–1978	Logarithm tables, slide rules	
1978–1996	Scientific calculators	
1997	Scientific calculators	Graphics calculators permitted
1998–2005	Graphics calculators	
2006–2009	Examination 1: No technology Examination 2: Graphics calculators	
2010–	Examination 1: No technology Examination 2: CAS calculators or software	New subjects aligning with Australian Curriculum from 2016 with unchanged technology.

in the three Year 12 mathematics subjects in Victoria (Australia) and also the technology, which examiners have assumed that students will be able to access when answering their questions. The progression is from scientific calculators nearly 40 years ago, to graphics calculators (programable, with a sizeable screen to observe and manipulate graphs, and substantial statistics functionality) and then to CAS calculators or software, which have all the functionality of the graphics calculator with an additional symbolic algebra (and calculus) functionality. Some of the machines also have dynamic geometry, but this is not relevant to the content of the Year 12 examinations. In this chapter, CAS refers to the capability of the software and not to whether it is implemented on a handheld "calculator" or a computer. In nearly all cases, technology has been permitted before its use has been assumed by the examiners. In the transition "permitted but not assumed" stages, questions were carefully designed so as not to disadvantage students without it.

The three subjects in Victoria have different content and are also directed to somewhat different student populations and are often referred to as *elementary, intermediate,* and *advanced* Year 12 mathematics. Because results in these subjects are used to determine university entrance and school performance, they are the single most important influence on technology use for mathematics in the local educational system. Victoria was the first Australian state to permit the use of graphics calculators, and then CAS in Year 12 state examinations. Whilst none of these changes has been uncontroversial, the adoption of CAS has been the most contentious and the most challenging for teachers. The sophistication of CAS software is the source of most of the challenge, whilst its capacity to "do" algebra and calculus is the source of most of the contention.

Further Mathematics, the elementary subject (although still at a Year 12 level), has the least emphasis on algebra, and consequently the decision to permit (but not assume) CAS calculators has been hardly controversial. The algebra required in that examination includes writing and interpreting equations of lines of best fit to data, writing recurrence relations, using formulas and solving equations in context. Perhaps the main reasons to permit CAS have been practical—some of these students will have used a CAS calculator in Year 11 and they should not have to make a special purchase for Year 12 or learn to use a different (graphics) calculator. These students mostly use digital technology for numerical calculation, graphing, and statistics.

The decisions to permit and later assume graphics calculators in the intermediate and advanced level subjects (*Mathematical Methods* and *Specialist Mathematics*) in 1997 generally proceeded smoothly despite generating some controversy (e.g., possible impact on students of higher cost of equipment; expected loss by some teachers of some by-hand mathematical skills such as sketching graphs; need for teachers' professional development). Many teachers quickly saw the advantage for teaching and learning of having easy access to manipulable graphs in these functions-and-calculus oriented courses. Some adjustments to examination questions were needed (see below). On the whole, however, the use of graphics calculators in class and in examinations has been embraced by teachers.

A series of studies demonstrated, however, that adopting CAS in examinations would have more impact. For example, Flynn & McCrae (2001) reported that questions producing 40% of the marks in the examination would be severely impacted if CAS was available rather than a graphics calculator. This was just one indication that moving to permit or assume CAS in Year 12 examinations (which automatically has the effect of encouraging it in teaching) would be more controversial, and would require greater adaptation of existing curriculum, assessment, and teaching practices. For this reason, from 2000–2002, a research project was established to study the curriculum, assessment, and teaching issues, and to introduce a new subject for examination, Mathematical Methods CAS (MMCAS), to run in parallel to the large and established graphics-calculator-active subject Mathematical Methods (MM). Many of the experiences in this pilot were reported by Stacey (2010) in the proceedings of the 2008 CSMC conference. The content of the subject MMCAS was essentially the same as MM, although it included working with a wider range of functions (e.g., the absolute value function was included because it so frequently appears in CAS results) and a few additional topics (e.g., transition matrices) to use time that might otherwise be allocated to perfecting by-hand skills.

Since the pilot study, the use of CAS has gradually become more widespread in schools. This has happened at Year 12, as documented in Table 17.1, but it is also worthwhile noting that it has driven interest in using CAS from Year 9 and above. See, for example, Pierce & Stacey (2011, 2013). Part of this interest is practical (students might only purchase one device whilst at school), part is because teachers expect that students' understanding might benefit, and part of it is a desire to maximize the exposure that students have to the practices that are relevant to the Year 12 examination. In the earlier years, the use is primarily pedagogical rather than computational, as students are still learning elementary algebra.

As Table 17.1 shows, the pilot MMCAS subject became available to all schools between 2002 and 2010, and finally the two subjects were merged once again with CAS as the assumed technology in 2010. An important modification during this period was the introduction of an assessment component with no permitted technology from 2006. The change was driven by fears that students would lose by-hand skills, and this is closely connected with a sense that understanding is developed and shown through by-hand skills. This solution has been adopted in various jurisdictions around the world. Drijvers (2009) says it is becoming the most common model in Europe. Other systems (e.g., College Board Advanced Placement Calculus AB and BC Program) permit students to have CAS but examiners do not assume it (College Board 2014; Leigh-Lancaster, Les, & Evans, 2010). In Victoria, the CAS active Examination 2 has 2 hours of testing time, whilst the technology-free Examination 1 has 1 hour of testing time and hence contributes one third of the weight of the external assessment to the students' overall mathematics grade. This arrangement of running closely parallel standard and pilot subjects has allowed a large-scale study of the effect of CAS on students' achievement, which is probably unique in the world. The results are described below. The most advanced subject, Specialist Mathematics, has had the same assumed technology (CAS) and examination structure as Mathematical Methods since 2010. Before this time, only graphics calculators were permitted (and assumed). Again practical issues (e.g., students only purchasing and learning to use one calculator) probably dominated the change, but the presence of the no-technology examination alleviated the still widespread concern that otherwise students might only be taught or only learn to press buttons without "understanding." Stacey & Wiliam (2013) provide a broad discussion of this and other issues related to assessment with mathematically able software. In all of these subjects, there is little use of technology other than the devices that students will use in the examination.

Changes to Examination Questions

Each of the changes in permitted and assumed technology has required changes to be made to examination questions. The introduction of the graphics calculator directly affected graphing/sketching functions. In addition, the capacity to solve equations numerically (by zooming in on graphs, using function tables or automatic "intersect" features) tended to orient problem solving in this way, and was to some extent counteracted by examiners frequently specifying that only exact answers would be acceptable. Figure 17.1 shows some questions that are fictitious but illustrate the type of examination questions used in the Mathematics Method subject and the changes that have occurred.

The demand of Question 1 is seriously affected when graphics calculators are permitted or assumed, since automatically graphing the function of the (hand calculated) derivative makes the question much simpler. For the graphics calculators or CAS eras, the question might be adapted as is shown in Question 2. Because the function rule is not given, the item more directly tests the knowledge that at the stationary points of a function the derivative is zero and hence stationary points match the x-intercepts of the derivative. Before graphics calculators, Question 3 was challenging: differentiate the function and solve the resulting cubic equation. With graphics calculators, this question became trivial—graph the function and zoom in to read the co-ordinates of the stationary points (see Figure 17.2).

Later graphics calculators (and today's models) give maxima and minima automatically. The "fix" for these questions was to ask for exact answers as in Question 4. However, this fix fails with CAS. With CAS, the question may still be worth asking but it is very much simpler, requiring only accurate entry and knowledge that the stationary points are the zeroes of the derivative function (see Figure 17.2). This illustrates how allowing CAS in examinations affects the demand of many more questions than allowing graphics calculators. Moreover it challenges the heart of what many teachers see as the main goal of learning the functions and calculus course—to be able to differentiate and integrate functions.

Before graphics calculators	**QUESTION 1** Which of the following four graphs below shows the derivative function of $f(x) = x(x-1)(x+1)(x-2)$? 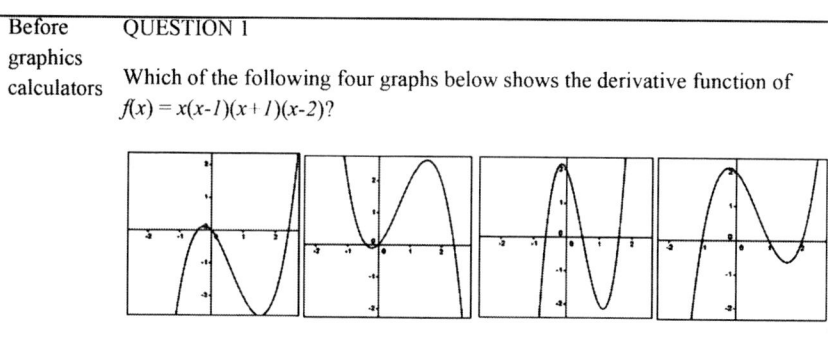
With graphics calculators or CAS	**QUESTION 2** Which of the four graphs (same options as above) shows the derivative of the function in this sketch? 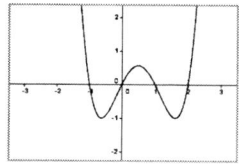
Before graphics calculators	**QUESTION 3** Find the stationary points of the function $f(x) = x(x-1)(x+1)(x-2)$.
With graphics calculators but before CAS	**QUESTION 4** Find exact stationary points of the function $f(x) = x(x-1)(x+1)(x-2)$ ANS: $x = (\sqrt{5}+1)/2$, $x = (-\sqrt{5}+1)/2$, $x = \frac{1}{2}$

Figure 17.1 Four questions illustrating changes to calculus questions.

EXAMINING STUDENT PERFORMANCE

Comparing Student Performance With and Without CAS

Having the two Mathematical Methods subjects with very similar content and different assumed and permitted digital technologies has provided a special context in which to study the effect of the algebraic aspects of CAS

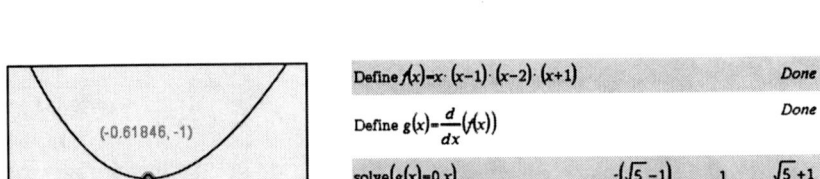

Figure 17.2 Finding stationary points by zooming in on a graph or with computer algebra.

on students' performance. In the first years of the parallel subjects, the students came from only a few schools, and, hence, there was uncertainty about the comparability of the two populations. However, in the last years of the two-subject arrangement, the size and representativeness of the two cohorts became comparable. Examination 1 (the technology free component) was identical and about three-quarters of the questions and marks allocated in Examination 2 were for identical questions, although potentially solved using different technologies. A series of studies by the examination authority's manager for mathematics (David Leigh-Lancaster) and its statistics experts and the chief examiners of the various subjects have explored the effect on student performance. These are summarized below.

Do Students Using CAS Lose By-Hand Skills?

A major concern for teachers and for many of the examiners is whether CAS students have poorer by-hand skills than students using graphics calculators, especially in algebra. The data for the no-technology Examination 1 for 2006–2009 all indicate that the CAS students perform at least as well as the graphics calculator students. Of course, judging by the examiners' reports that describe student responses to each question, the skills of both groups remain below what the examiners desire! But this has been the case throughout living memory! The best comparison data are from 2009, the last year of two separate subjects, when the MMCAS cohort had grown to be nearly the same size as the MM cohort (7189 MMCAS and 8887 MM students). Leigh-Lancaster et al. (2010) reported that there were similar percentages of students getting excellent scores in the two groups, that fewer CAS students received very low scores, and that the mean score of the other (middle) students was slightly higher for the CAS students. This pattern persists when the mathematics, science, and technology component of the general ability test (completed by all students) or the overall score for all

Year 12 subjects (English, languages, sciences, etc.) is used as a covariate to control for possible bias in the selection of students for MMCAS. There is, therefore, no evidence that the populations of students have different abilities. Choice of which subject (with or without CAS) to undertake is a decision of the school (generally the Year 12 mathematics teachers), not of individual students.

How Does Using a CAS Rather Than a Graphics Calculator Affect Performance?

Examination 2 in the years 2006–2009 was done with technology: graphics calculator for MM and CAS calculator (and in a few instances computer software) for MMCAS. Because of the very large overlap of content, many of the questions are the same, enabling comparison of students' learning and performing with or without access to symbolic algebra. A consistent pattern of slightly better performance is found in the total scores on the common questions in Examination 2 (about half of the examination), both when analyzing the scores alone and when using the mathematics, science, and technology component of the general ability test as a covariate. There have been several different analyses, and data from every year have been examined. Norton, Leigh-Lancaster, Jones, and Evans (2007) looked at the questions in Examination 2 that were common to the two subjects. They found that, in the 2006 examinations, MMCAS students generally performed better overall than MM students both on common multiple-choice questions and on common parts of extended response questions. In fact, in 2006 there were only 2 of the 35 common questions where the MM cohort performed statistically better than the MMCAS cohort whilst there were 12 where the MMCAS cohort was better.

Following several other researchers (e.g., Flynn and McCrae [2001]), common questions were classified by the researchers as technology-independent (e.g., finding the maximal domain of $log|x-b|$); technology of assistance but neutral with respect to graphics calculators or CAS (e.g., finding the numerical probability of 8 or more heads in tossing a coin 10 times); or technology-advantageous when use of CAS is likely to make the question simpler (e.g., solving $|2k + 1| = k + 1$).

Questions where the MMCAS group performed better were in all three categories. Norton et al. (2007) commented that one advantage of using CAS is that, once a solution method has been formulated, it is often simple to carry out the method using CAS, thus avoiding trivial algebraic errors. This then allows the student with CAS to engage easily and more confidently with further parts of the question. The finding of the same or better performance on common questions was replicated in all subsequent years until the two subjects merged.

Leigh-Lancaster et al. (2010) applied item response theory (a two-dimensional Rasch regression model) to the 2009 scores to compare (separately) the scores of students on the multiple choice and the extended answer questions of Examination 2 that were in common for the two subjects. The common questions made up the majority of the examination: 17 of 22 multiple-choice questions, and 21 of 32 extended answer questions. In this analysis, questions in the third category (where CAS was expected to have an advantage) were not regarded as being in common. When using the students' score on Examination 1 as a measure of ability (recall that both cohorts did the same no-technology assessment), it was found that MMCAS students performed slightly better than MM students across the ability range. The only exception was that performance of the best students was similar. The peak difference for the multiple-choice questions occurred with the below average students (score around –1 logits using item response theory), whereas the peak difference for the extended-response questions was for students scoring a little better than average (around 0.5 logits) on Examination 1. These students are evidently able to use the algebra functionality to advantage, either to solve problems completely or to check their work.

Why Do the MMCAS Students Perform Slightly Better on Common Questions?

Results on common questions for the MMCAS cohort are slightly better than the results for the MM cohort, even when controlled for "ability." Why is this? The examiners put forward several reasons. As noted above, students might have avoided trivial algebraic errors, and this enabled them to continue with further parts of the question. Evans, Norton, and Leigh-Lancaster (2005) note that scrutiny of CAS examination scripts over three years showed that "CAS does 'scaffold' and enable students to engage, and continue to engage, in extended-response analysis questions, with comparatively good level of success" (pp. 333, 334). They also reported several instances where the MMCAS group was more accurate than the MM group, even though both technologies were useful. For example, in 2004, 90% of MMCAS students and 80% of MM students correctly found a numerical solution to the equation $ln(x + 1) = 1 - x$ correct to two decimal places. Evans et al. postulated that the requirement to enter symbolic expressions very carefully into CAS encourages better attention to detail. Many errors with graphics calculators relate to errors of input. At the time that Evans et al. drew this conclusion, it is unlikely that the CAS calculators had an easier interface than the graphics calculator—possibly the reverse—so it seems to lie within the students' practice rather than the technology.

Analysis shows that the positive results are not due to individual student ability. However, might the above results instead be due to bias in the sample of schools (and hence teachers) that moved to CAS early in the

2000–2010 period, rather than be due to the technology used? Possibly students at schools that changed to MMCAS earlier were of a slightly higher socio-economic status than those who were slower to change, but the cost of the CAS calculators was comparable to the cost of graphics calculators (except initially), and ability has been accounted for. Possibly teachers in schools choosing the new subject MMCAS over the traditional one were more mathematically adventurous, and, hence, somehow better teachers. There are no data to decide this hypothesis, although it is interesting to note that some of the very highest performing schools tended to be slow to change (Leigh-Lancaster et al., 2008), perhaps anxious to preserve their reputation. However, by 2009, the last year of parallel subjects, both cohorts were large and the effects persisted.

In their explanation of the results, these research teams also included the possibility that better learning was enabled by the symbolic functionality of CAS. The graphics calculators already enabled moving smoothly between numerical and graphical representations of functions, but the symbolic algebra extends this considerably. Symbolic algebra also supports student exploration of algebraic properties such as the calculus product rule (Leigh-Lancaster, 2010). The international literature on teaching with CAS provides many more examples of how learning can be positively affected. However, it is important to point out that the studies reported here of students' examination scripts and scores are consistent with there having been a learning benefit, but these data do not provide direct evidence that deeper conceptualization and other predicted learning benefits have occurred. Studies with other methodologies do this.

Gender Equity

There were some early reports (see, for example, Forgasz & Tan, 2010) that using CAS rather than graphics calculators disadvantages girls. In 2009, both cohorts were large and there was no reason to expect any bias in the sample related to gender, although strangely the proportion of girls in the two subjects was different (42% for MM and 46% for MMCAS). The choice of subject would have nearly always been made by the school, so this probably indicates different choices by single sex schools. One indicator of a changing gender difference is to look at the number of students being awarded the top A or A+ grades, because these grades are where the differences in favor of boys are usually more evident. (This is a general phenomenon of mathematics assessment, by no means confined to these subjects). Using data from HREF3, in Exam 1, the percentage of MM males scoring A or A+ was 1.65% more than for females, whereas the difference in percentages was 3.21% for MMCAS. For Exam 2 the gaps were a little higher: 3.1% more males got A or A+ in MM and 4.04% more males got A in MMCAS. It is unexplained why the A grade gender gap is larger in MMCAS than MM.

However, these data do not obviously support a hypothesis that using the CAS disadvantages girls over using graphics calculators, because the gender gap is similar for Examinations 1 and 2, with and without technology. It is well established that there are gender differences in the on-average responses of boys and girls to surveys about using technology for mathematics. For example, Pierce, Stacey, and Barkatsas (2007) confirmed a general result that boys tended to positively enjoy using technology for mathematics, whereas girls tended to value it only for its use in solving their allocated problems. However, an argument for increased disadvantage for CAS needs to explain why girls do not learn by-hand skills as well as boys in a CAS environment. In my judgement, it is unlikely that putting the additional algebra capability of the CAS in the hands of girls is responsible for this gender gap.

Calculators or Computers: Comparing Student Performance

With the increasing range, availability, and power of mobile computing devices, there is some pressure from schools for students to use CAS on a computer (or tablet device) instead of having to purchase a dedicated calculator. Assessment using computers instead of pen-and-paper is also of interest to examination bodies for all subjects, not just mathematics. Consequently, there has been a trial using computers for all aspects of the Mathematical Methods examination process—question delivery with text and images, students' computations and answers, access to the supplied formula sheet, and later also the scoring.

Zoanetti, Les, and Leigh-Lancaster (2014) report a study where there were 62 volunteer students from five schools who studied the same curriculum and sat the same examinations under the same general conditions (apart from the mode of delivery and response) as the standard cohort. Examination 1 (no technology) was the same. For Examination 2, the trial students used Mathematica Notebooks on a computer and the standard cohort used a CAS calculator and handwrote their responses.

There were no statistically significant effects of the mode of assessment. For example, mode did not change the prediction of the Examination 2 score from the Examination 1 (no technology) score, or from the Examination 1 score and the students' scores on the mathematics, science, and technology component of the general achievement test undertaken by all students presenting for the end-of-school certificate. Only one of the 22 multiple choice questions showed statistically significant differential item functioning (DIF): a question requiring students to identify that a certain variable was normally distributed and then find an associated probability. There is no a priori reason why this item should be affected by using a computer rather than a hand-held calculator. This study gives confidence that the newly designed assessment procedures were fair, although further monitoring is needed as the number of students and number of devices grows.

AUSTRALIAN CURRICULUM

Since the beginning of compulsory education, each Australian state has had what might be called its own "national curriculum," with statements of what is to be learned, and culminating in state-organized examinations used for end of school certification and university entrance. At the present time, all the states (including Victoria) are moving towards a common "Australian curriculum" (http://www.australiancurriculum.edu.au/), although state control of the end of school examinations remains. Perhaps in order to accommodate state differences in the preference for use of technology, the Australian mathematics curriculum does not specify particular software or hardware, using statements such as "Access to technology to support the computational aspects of these topics is assumed" (HREF1). Use of spreadsheets is also explicitly mentioned in the science curriculum and in the section on consumer mathematics. The description of the generic capability with ICT for mathematics includes learning to use it effectively and appropriately "to perform calculations, draw graphs, collect, manage, analyze, and interpret data; share and exchange information and ideas and investigate and model concepts and relationships" (HREF2). There is also recognition of pedagogical potential: "Digital technologies, such as spreadsheets, dynamic geometry software and computer algebra software, [to] engage students and promote understanding of key concepts" (HREF2). The latter is the only mention I found of computer algebra in the Australian curriculum. Despite this, the new courses in Victoria, created in alignment with the Australian curriculum for adoption from 2016, retain the use of computer algebra as above.

WHAT USE IS BEING MADE OF CAS IN SCHOOLS?

Since the 2000–2002 research project, when the small group of research project teachers' use of CAS was carefully documented, there have been only a few studies of CAS use in class. Pierce and Ball (2009) reported a large survey of the perceptions about technology use of teachers in Victoria. They found that teachers were mostly optimistic about effects of using technology on teaching and on student learning. However, within the sample, 25% were strongly concerned about effect on by-hand algebra skills and 25% felt that the need to teach CAS use takes too much time from effective mathematics instruction. These groups of dissenting voices overlap and include many mid-career teachers (rather than the older or younger teachers).

Pierce and Stacey (2013) studied "early majority" adopters of computer algebra, who were Year 9 and 10 teachers embarked on a multi-year professional development project organized by their school. The results underlined the facts that many teachers find it very hard to learn to use the technology well and sufficiently fluently to use it confidently in class, and that advances in technology mean that they need to keep learning. It is not just the CAS, and its rather different way of thinking about mathematics, but also the associated presentation technologies (e.g., interactive whiteboards) that add to the total load. The transition in teaching practices can be extremely slow. Above all, despite the capabilities of the technology, there is a strong tendency to use the CAS just like a "calculator"—providing step-by-step assistance to unchanged by-hand work. So, for example (although not from that study), the calculation of the stationary points in Figure 17.2, where the actual derivative is not shown, may be felt to be very unsatisfactory. As another example, we have seen teachers new to CAS advise students to use the product rule for differentiating a product, using the CAS to differentiate only the individual factors rather than the complete product and putting together the resulting expressions by hand. The power of CAS is not evident on first glance and new mathematical practices are slow to arise and gain legitimacy.

Sue Garner, one of the teachers involved in the original research project, has reviewed her experiences of teaching with CAS since 2001 (Garner, 2014). She describes how some teachers with whom she has worked closely have encouraged students to use CAS algebra really only to check answers. They avoid the "explosion of methods" that occurs when using technologies with multiple representations and high power. In contrast, with some colleagues, she reports that new norms for teaching have been established, especially using what she calls "teaching the ends and sides of a topic." Garner reports success turning teaching around, to start with purpose and applications, and then look at techniques in detail. She enjoys celebrating the explosion of methods by getting more student input. Many students find this very good, although they acknowledge: "[CAS] is a friend that is useful sometimes, annoying sometimes, and plainly a waste of time at other times. The task is to find when!" (a quote typical of successful students).

Garner has classified her students as *stayers*, who steadily achieve in all environments; *resenters* who feel that CAS devalues their skills; *flyers*, who find and enjoy new solution methods, relationships, and patterns with CAS; and *enabled* for whom CAS can compensate for unreliable by-hand algebra. Whereas her early classes of CAS students had small but significant numbers of enabled students, Garner reports that these students are no longer present, perhaps because of the existence of the no-technology examination component or modifications to questions to "beat the CAS." The 2012 examination question in Figure 17.3 illustrates this. Before CAS a similar question might have been asked about a single point—to show that at the

> If (p, q) is any point on the graph of $y = f(x)$, where $f(x) = \dfrac{1}{2x-4} + 3$, show that the equation of the tangent to $y = f(x)$ at (p, q) can be written as $(2p - 4)^2(y - 3) = -2x + 4p - 4$.

Figure 17.3 VCAA Mathematical Methods examination question 2012.

point $(3, \tfrac{1}{2})$, the tangent line is $y = -\tfrac{x}{2} + 5$. With the 2012 question, CAS gives the tangent line as

$$(y =) \frac{6p^2 - x - 22*p + 22}{2(p-2)^2},$$

but the question requires a different algebraic form. It appears that by-hand algebra is required to get full credit.

An example from the technology-permitted examination of the "advanced" subject Specialist Mathematics is shown in Figure 17.4. CAS can readily solve the differential equation (although the decimals make it all a bit messy!), but part (a) instead asks for an identity that is not used in any later parts to reveal anything about the solution. It seems that this part of the question, despite being in the technology-permitted examination, is largely motivated by the desire to test by-hand algebra.

A recent survey carried out by Pierce and Bardini (2015) has tried to gather some information on a wider group of teachers' CAS use. They surveyed first year mathematics students at the University of Melbourne, an above average cohort in 2013/2014. Of 2000 first-year students, 334 answered the survey and had used CAS in their end-of-school examinations. Students were asked about their own use of CAS and their perceptions of teachers' use of CAS in class for pedagogical purposes, such as scaffolding the learning of pen-and-paper skills, exploring regularity and variation, linking representations, etc. They found that the students reported less

> **Question 3** (11 marks)
> The number of mobile phones, N, owned in a certain community after t years, may be modelled by $\log_e(N) = 6 - 3e^{-0.4t}$, $t \geq 0$.
> a. Verify by substitution that $\log_e(N) = 6 - 3e^{-0.4t}$ satisfies the differential equation
> $$\frac{1}{N}\frac{dN}{dt} + 0.4\log_e(N) - 2.4 = 0.$$
> 2 marks

Figure 17.4 Question 3 from Examination 2, Specialist Mathematics 2013, Victorian Curriculum and Assessment Authority.

in-class pedagogical use by teachers than might have been expected. There was a lot of use of graphing, but little pedagogical use for scaffolding learning algebra or for extending algebraic demands of problems. A few teachers used CAS often in most topics, but the majority was skewed towards little in-class demonstrated use.

ISSUES ARISING

It is now over 30 years since Wilf (1982) wrote his article "The disk with the college education," in which he described the capabilities of the computer algebra system "mµmath," which was the first to run on a home computer. He raised the question of how secondary and college mathematics curriculum, assessment, and teaching should change when such software and hardware were readily accessible to all, perhaps in the form of a $29.95 calculator (in 1982 prices!).

In Victoria, and elsewhere around the world, it has now been established that this mathematical able software can be used to pedagogical advantage, and that it can also be incorporated as a fundamental tool for learning and doing, teaching, and assessing mathematics. Teachers and students are able to learn how to use it and to teach with it. Many, although not all, find it rewarding to use, adding new dimensions to mathematical learning. Assessment can be equitable. The Victorian experience has also shown that, even within a large and diverse system, the basic algebraic skills of students who learn with computer algebra technology can be maintained at a level comparable to that of students who do not use it. The dual assessment model (part with, part without technology) seems to have strong support from teachers and has emerged as a standard model around the world. All of these are successes.

However, some old challenges remain unresolved, and there are some new challenges looming ahead. The main longstanding challenge is deeply connected to what people think mathematics is and what they most value. Wilf concluded his 1982 article wondering whether teachers of more advanced mathematics would take the advice that was frequently given to elementary school teachers concerned about the impact of four-function calculators: that they should concentrate on the concepts rather than the mechanics of mathematics. This has certainly been done in some parts of the mathematics curriculum. For example, it is hard to find in the literature any arguments against using technology for the by-hand processing of statistics in either teaching or assessment (Stacey & Wiliam, 2013). However, little progress has been made in separating the concepts from the mechanics in algebra. Some evidence for this is: (a) the popularity of the no-technology examinations, (b) instructions to show working even when technology is permitted, and (c) the way in which many examinations tend to include questions requiring algebraic forms that are hard to squeeze from the technology, as noted above.

Perhaps the greatest unresolved issue is how to use technology as an amplifier rather than to compensate for inadequate algebraic skills. The new curricula have generally not moved strongly in the direction of using technology to do more—for example to put the focus on mathematical modeling of real world problems rather than on intra-mathematical calculation. Brown (2010) surveyed six examination systems and found this to be the case. He attributes it to the examiners' views of what is most important in mathematics, although other factors may be that new skills are required to write good questions for the new environment and that moving away from the testing of "mechanical" aspects of mathematics may make it too difficult for many students (Stacey & Wiliam, 2013).

After analyzing the mathematical literacy required in industry and business to respond to the new data-rich, visualization-rich, and computationally-rich environment, Hoyles, Noss, Kent, & Bakker (2010) coined the term "techno-mathematical literacies" to describe the inter-dependence of mathematical literacy and the use of information technology for employees at all levels in the workplace. Using specialized workplace systems and also open mathematical tools—especially for statistics, graphing, data handling, three-dimensional visualization, and algebra—requires both the understanding of the underlying mathematics as well as being able to think in the ways that using the technology demands and considering the workplace context. Should schools teach these potentially work-place specific techno-mathematical literacies, and, if so, how?

New challenges come from ever-changing forms of technology and new developments in mathematics. As a living subject and with the impetus of being able to do very much more with new technology, mathematics, and the way it is used are changing. Are there new topics that should be introduced (e.g., handling "big data") and, if so, what old topics should move out of school mathematics to make way for them? There is a sense that the mathematics curriculum content needs to change, but in what direction? The inclusion of statistics over recent decades has been very significant in this regard, but what comes next?

The challenges from new forms of technology for computation will become more pressing for teachers and assessment systems. Up until now, teachers have decided what technology type and even brand their students will have, considering the hardware and software that are permitted in the significant assessment systems. In Victoria, this means that generally all students in the class have the same hardware. Teachers instruct their students how to use this one device. This seems likely to change. Schools may adopt "bring your own device" policies (see, for example, Government of Australia [2013]) and not all of these will support the same mathematics software. Free on-line services, such as Wolfram Alpha, have considerable mathematical power, including step-by-step solutions, and may appear more attractive than dedicated devices. A wide range of special "solvers" are now easy to find on the Internet. Commonly used examples are currency conversion apps and the TVM solvers for the time value of money (for calculating load repayments, periods, etc.) and frequently a "wizard" will guide you through their use. Special purpose apps like this could also be embedded into textbooks and other learning materials. What are effective pedagogies for teaching with this diversity? Using solvers of this type fits into the general way in which technology is used at school like a calculator in support of line-by-line by-hand reasoning. But there are other possibilities, including multi-page open tools that combine representations and functionality to support working on complete problems. Does doing mathematics with technology mean mastering an open general purpose tool that deals with many branches of mathematics, or could it mean being able to use many special purpose apps? What mathematical understanding is needed for this?

Mathematics teaching has encountered practical and philosophical challenges from digital technology since the 1970s. There has been a lot of progress, and certainly the use of CAS in Victoria must be regarded as a well-accepted, firmly established and successful innovation. However, some of the philosophical challenges are not fully resolved and new practical challenges will continue to arise as the digital environment continues to change.

REFERENCES

Brown, R. (2010). Does the introduction of the graphics calculator into system-wide examinations lead to change in the types of mathematical skills tested? *Educational Studies in Mathematics, 73*(2), 181–203.

College Board. (2014). *AP Calculus AB (Calculator Policy)*. Retrieved from https://apstudent.collegeboard.org/apcourse/ap-calculus-ab/calculator-policy

Drijvers, P. (2009). Tools and tests: Technology in national final mathematics examinations. In C. Winslow (Ed.), *Nordic research on mathematics education: Proceedings from NORMA08* (pp. 225–236). Rotterdam, Netherlands: Sense.

Evans, M., Norton, P., & Leigh-Lancaster, D. (2005). Mathematical methods computer algebra system (CAS) 2004 Pilot Examinations and links to a broader research agenda. *Proceedings of 28th Conference of the Mathematics Education Research Group of Australasia* (pp. 329–336). Retrieved from http://www.merga.net.au/documents/RP342005.pdf

Flynn, P., & McCrae, B. (2001). Issues in assessing the impact of CAS on mathematics examinations. *Proceedings of 24th Conference of the Mathematics Education Research Group of Australasia* (pp. 222–230). Retrieved from http://www.merga.net.au/documents/RR_Flynn&McCrae.pdf

Forgasz, H., & Tan, H. (2010). Does CAS use disadvantage girls in VCE Mathematics? *Australian Senior Mathematics Journal, 24*(1), 25–36.

Garner, S. (2014, November). *CAS: Ten years on.* Paper presented at the National Council of Teachers of Mathematics Regional Conference, Houston, Texas.

Government of Australia. Digital Education Advisory Group. (2013). *Beyond the classroom: A new digital education for young Australians in the 21st century.* Retrieved from http://apo.org.au/research/beyond-classroom-new-digital-education-young-australians-21st-century

Hoyles, C., Noss, R., Kent, P., & Bakker, A. (2010). *Improving mathematics at work: The need for techno-mathematical literacies.* London, England: Routledge.

HREF1 http://www.australiancurriculum.edu.au/seniorsecondary/mathematics/specialist-mathematics/curriculum/seniorsecondary#page=1

HREF2 http://www.australiancurriculum.edu.au/mathematics/general-capabilities

HREF3 http://www.vcaa.vic.edu.au/Pages/vce/statistics/2009/statssect3.aspx#H3N10261

Leigh-Lancaster, D. (2010). The case of technology in senior secondary mathematics: Curriculum and assessment congruence? *Proceedings of 2010 ACER research conference* (pp. 43–46). http://research.acer.edu.au/cgi/viewcontent.cgi?article=1094&context=research_conference

Leigh-Lancaster, D., Les, M., & Evans, M. (2010). Examinations in the final year of transition to mathematical methods computer algebra system (CAS). *Proceedings of 33rd Conference of the Mathematics Education Research Group of Australasia* (pp. 336–343).http://www.merga.net.au/documents/MERGA33_Leigh-Lancaster&Les&Evans.pdf

Leigh-Lancaster, D., Norton, P., Jones, P., Les, M., Evans M., & Wu, M. (2008). The 2007 common technology free examination for Victorian certificate of education (VCE) mathematical methods and mathematical methods computer algebra system (CAS). *Proceedings of 31st Conference of the Mathematics Education Research Group of Australasia* (pp. 331–336). Retrieved from http://www.merga.net.au/documents/RP382008.pdf

Norton, P., Leigh-Lancaster, D., Jones, P., & Evans, M. (2007). Mathematical methods and mathematical methods computer algebra system (CAS) 2006–Concurrent implementation with a common technology free examination. *Proceedings of 30th Conference of the Mathematics Education Research Group of Australasia* (pp. 543–550). Retrieved from http://www.merga.net.au/documents/RP492007.pdf

Pierce, R., & Ball, L. (2009). Perceptions that may affect teachers' intention to use technology in secondary mathematics classes. *Educational Studies in Mathematics, 71,* 299–317.

Pierce, R. & Bardini, C. (2015). Computer algebra systems: Permitted but are they used? *Australian Senior Mathematics Journal, 29*(1) 32–42.

Pierce, R., & Stacey, K. (2011). Lesson study for professional development and research. *Journal of Science and Mathematics Education in Southeast Asia, 34*(1), 26–46.

Pierce, R., & Stacey, K. (2013). Teaching with new technology: Four "early majority" teachers. *Journal of Mathematics Teacher Education, 16*(5), 323–347.

Pierce, R., Stacey, K., & Barkatsas, A. (2007). A scale for monitoring students' attitudes to learning mathematics with technology. *Computers and Education, 48*(2), 285–300.

Stacey, K. (2010). CAS and the future of the algebra curriculum. In Z. Usiskin, K. Andersen, & N. Zotto (Eds.), *Future curricular trends in school algebra and geometry* (pp. 93–108). Charlotte, NC: Information Age.

Stacey, K., & Wiliam, D. (2013). Technology and assessment in mathematics. In M. A. (Ken) Clements, A. Bishop, C. Keitel, J. Kilpatrick, and F. Leung (Eds.), *Third international handbook of mathematics education* (pp. 721–752). Dordrecht, Netherlands: Springer.

Wilf, H. (1982). The disk with the college education. *American Mathematical Monthly, 89*(1), 4–7.

Zoanetti, N., Les, M., & Leigh-Lancaster, D. (2014). Comparing the score distribution of a trial computer-based examination cohort with that of the standard paper-based examination cohort. *Proceedings of 37th Conference of the Mathematics Education Research Group of Australasia* (pp. 685–692). Retrieved from http://www.merga.net.au/documents/merga37_zoanetti.pdf

CHAPTER 18

MATHEMATICS EDUCATION IS AT A MAJOR TURNING POINT

David Moursund

> *If you want to teach people a new way of thinking, don't bother trying to teach them. Instead give them a tool, the use of which will lead to new ways of thinking.*
> —Richard Buckminster Fuller, American engineer, author, designer, inventor, and futurist, 1895–1983

> *Don't worry about what anybody else is going to do... The best way to predict the future is to invent it. Really smart people with reasonable funding can do just about anything that doesn't violate too many of Newton's Laws.*
> —Alan Kay, American computer scientist, 1940–[1]

The quotations from Buckminster Fuller and Alan Kay capture the essence of this chapter about possible futures of mathematics education. My goal is to help invent a better mathematics education system.

We now have artificially intelligent computer brains, and these are steadily growing in capability. We know that human brains and computer brains working together can accomplish a wider variety of problem-solving tasks than either working alone. This idea that the educational use of the two brains, human and computer, is better than using only one or the other alone underlies much of my teaching and writing (Moursund, 2012, 2013, 2014a, 2014e).

Some progress is occurring in applying this two-brain capacity to improving our educational system. The two-brain approach is steadily being improved to help people represent and solve math-related problems, to learn mathematics, and to teach mathematics. Although this chapter deals primarily with mathematics education, its basic ideas apply to education in all disciplines of study and research. I hope some of the future-oriented ideas presented in this chapter will resonate with readers, and that together we will succeed in improving our world's mathematics education systems.

TWO FUNDAMENTAL QUESTIONS

Over the years, some of my educational ideas and forecasts have proven reasonably correct, while others have missed the mark. For example, in the late 1960s and early 1970s, my understanding of computers as an aid to *doing and using mathematics* convinced me that the U.S. K–12 math curriculum would quickly adjust to using computers to represent and solve a wide variety of problems. Read more of my forecasts for the evolution of mathematics education in Moursund (1989).

Forty years have passed, and I am still waiting! Our mathematics education system has yet to adequately address two fundamental questions:

- If a computer can solve or greatly help in solving a category of mathematics problems for which we have traditionally taught "by hand" methods of solution, what should students be learning about solving such problems?
- What content currently in the curriculum should be downplayed or dropped, and what content should be added, as computer capabilities and availability continue to increase?

IMPROVING MATHEMATICS EDUCATION

It is easy to claim that our mathematics education system is not as good as it could and should be. It is much harder to actually develop and implement changes that will substantially improve the system. Implementation of any major change is difficult, as it can require substantial retraining of teachers and overcoming the resistance to change that parents and other major stakeholder groups often have.

However, I believe our mathematics education system is at a major turning point. We now have the knowledge and the technology to make a huge change for the better in this system. See "Improving Mathematics Education" (Moursund, 2014b) for an extensive treatment of my ideas for this improvement.

In this chapter, I present a number of ideas that I believe will collectively put our math education system on a new and much better track. Although there is current research to support some of these ideas, additional research is still needed. In the list given below, I briefly summarize some ideas/forecasts that I believe can facilitate substantial improvement in mathematics education. Computerized, intelligent teaching machines are emerging as a vehicle that will make all of these ideas possible.

1. Teaching machines that provide individualization of instruction and also make substantial use of academically and socially oriented computer networking will become a standard and routinely used aid to teaching and learning throughout the world.
2. Problem solving lies at the heart of goals for mathematics education. Computer technology is a powerful aid to problem solving in mathematics and many other disciplines, and will be thoroughly integrated into the mathematics curriculum (Moursund, 2014d).
3. Mathematics is a powerful aid to problem solving in many non-math disciplines. Computer technology is now routinely used to integrate the use of mathematics into the various curriculum areas.
4. Increasing students' mathematics maturity will be more accepted as a worthy and achievable mathematics education goal (Moursund, 2014c).
5. Teaching and learning will become more of a science. Brain science (cognitive neuroscience) is providing us with a much better understanding of how people learn (Bransford, Brown, & Cocking, 1999; Moursund, 2014a). Current online courses will increasingly be used to collect data on student performances that can be analyzed to improve the online systems (PACT Center, n.d.).

The term "teaching machine" as used in this chapter can be thought of as a modern and continually improving Information and Communication Technology tool for online and offline education. There has been considerable research on the effectiveness of online education. Thomas Russell's website, "The no significant difference phenomenon," contains a list of many hundreds of such studies (Russell, 2014).

THE TOOL IS THE TEACHER

The medium is the message.
—Herbert Marshall McLuhan, Canadian philosopher
of communication theory and a public intellectual, 1911–1980

Marshall McLuhan is well known for this statement. In applying his statement to the medium of computers in education, I have begun using the statement, "The tool is the teacher." I mean this in two senses.

1. Instruction on how to use a computerized tool can be built into the tool.
2. We can build a tool—a computerized general-purpose teaching machine—that will revolutionize our educational systems.

We have had computerized teaching machines since the late 1950s. The following example illustrates an early computer system that could both help solve an important problem and also could teach its users. It provides an excellent early example of the *tool is the teacher*.

Beginning in the late 1950s, the United States and Canada developed the North American Aerospace Defense Command (NORAD).[2] NORAD included an "early warning" system of radar and computers that could detect and report airplane and missile-based threats launched over the North Pole toward the United States and Canada. Operators viewed a computerized TV display screen and could act on the data they were receiving. The same display screen could show simulated (previously recorded or computer generated) data. Consequently, system operators could be trained/educated using quite authentic simulations. The integration of a problem-solving tool with a teaching tool was a breakthrough in teaching machines.

The 1985 science fiction book *Ender's Game* provides a much broader example of education that is based on the idea that the tool is the teacher (Card, 1985). In this story, a boy named Ender spends years learning how to play a computerized war game. We learn late in the story that playing the game is indistinguishable from using the tools built into the game to fight a "real" war.

Here is my own statement of an updated version of Buckminster Fuller's statement quoted at the beginning of this chapter: "If you want to teach people a new way of thinking and problem solving, give them a computerized tool that has the *intelligence* to teach them how to use the tool even as they are using the tool."

In the past, human tutoring and apprenticeships served as the intelligence that provided instruction and individualized feedback. What is new is that Highly Interactive Intelligent Computer-Assisted Learning (HIICAL) systems (Sylwester & Moursund, 2012) and Computer Tutor systems (PACT Center, n.d.) are becoming better and better in serving as this intelligence. In brief summary, a growing body of research and development in the areas of HIICAL, Computer Tutors, educational computer simulations, and other aids to teaching and learning is making it clear that they can play a significant role in improving educational systems.

COMPUTERS AND PROBLEM SOLVING

Perhaps the most important aim in improving mathematics education is to improve students' math-related problem-solving knowledge and skills. Mathematics educators have worked on this task for years. The Common Core State Standards for Mathematics (CCSSM) in the United States focuses on improving mathematics problem solving (National Governors Association Center for Best Practices & Council of Chief State School Officers, 2010). However, it seems to me that CCSSM is merely doing "more of the same," but with a shift to more emphasis on problem solving and understanding. I don't believe CCSSM contains any major breakthrough ideas that will substantially improve the mathematics education system.

Here is a simple example that illustrates the computers-in-math education challenge. The square root key on an inexpensive calculator provides us with a clear distinction between the task of calculating a square root and the task of understanding both the underlying math and some uses of the square root function. The math curriculum, as defined by widely sold textbooks, has made the transition to use of calculators. Interestingly, however, my recent Google search of "hand calculation of square root" produced about 7,390,000 results. (See, for example, Ask Dr. Math [n.d.].).

The teaching of statistics in higher education provides another useful example. Today's statistical packages of computer programs have removed the need for a user to have a deep understanding of the underlying mathematical theory. Most students who need to make use of statistics in their major fields of study take discipline-specific statistics courses that include a major focus on the use of computerized statistical packages. Such courses focus on when a particular statistical analysis might be suitable, and the possible meaning of the results rather than on calculation methods.

Computers have added a new general category to mathematics. We now have pure math, applied math, and computational math. In mathematics education, computer technology lets us decrease the amount of emphasis on mastering routine manipulative aspects of mathematics. It adds a new challenge of learning to do math-related computational thinking[3] and computer analysis of very large databases (big data). Computational mathematics can be taught in departments of mathematics, but also (like statistics) lends itself to being taught in a discipline-specific manner in many other departments.

Here is an example of how computers are changing problem solving.

Eric Chen, a High School Senior

The December 2013 issue of *Scientific American* included a story about Eric Chen, who was a 17-year-old high school senior from San Diego,

California, when he was the Grand Prize winner in the 2013 Google Science Fair (Kuchment, 2013, October). Quoting Chen:

> I live in San Diego, where some of the first cases of 2009 H1N1 swine flu took place in the U.S. It was then that I made a realization that flu can kill a lot of people. I thought, "Why can't we use the new computer power at our fingertips to speed up drug discovery and find new flu medicine?" I came across Dr. Rommie Amaro of the University of California, San Diego, and she was willing to let me work in her computation lab.

Chen then goes on to describe his activities of using the computer to screen a half-million chemical compounds, separating out 237 likely candidates, and testing each of them in a "wet" lab (that is, a "traditional" biology lab) to identify six that he deemed worthy of animal studies. With an inquisitive and persistent mental demeanor, some tutoring from a professor, and the help of computer technology, a high school student was able to do cutting edge research in medicine. What a marvelous learning experience and example of Buckminster Fuller's statement!

MATHEMATICS MATURITY

I think of mathematics education as having two broad goals: students learning *mathematics content* and gaining in *math maturity*. The term "math maturity" is often used in discussing a person's math-oriented general knowledge, skills, insights, ways of thinking, and habits of mind that develop and endure over the years. Most students quickly or gradually forget many of the details of the mathematics content they have studied but do not routinely use. However, students' success in increasing their level of mathematics maturity tends to serve them for a lifetime.

The new CCSSM acknowledge the importance of helping students develop math maturity. However, the term is not carefully defined in the CCSSM and receives little emphasis. Math maturity is not just something that one has or does not have. It also is not a specific component of the mathematics content that is taught in schools. Rather, one's level of math maturity grows through the study and use of mathematics. Here are some examples (certainly not an exhaustive list) that help to convey some basics of math maturity.

Math maturity includes the ability and/or capacity to:

- Understand the discipline of mathematics as a very important (and enjoyable to many) human endeavor.
- Develop math-oriented habits of mind.

- Learn to learn mathematics; making the significant shift from learning by memorization to learning through understanding.
- Recognize a valid mathematical/logical proof or argument, and detect "sloppy" thinking. Provide solid evidence (informal and formal arguments and proofs) of the correctness of one's problem-solving efforts.
- Pose and/or recognize mathematics problem situations and patterns of interest to oneself and others.
- Transfer one's mathematical knowledge and skills into math-related areas and problems in disciplines outside of mathematics.
- Understand the capabilities and limitations of tools (including calculators, computers, the Internet, and the Web) to help represent and solve problems. Learn to make effective use of these tools at a level commensurate with one's overall knowledge, skills, and understanding of mathematics.

Notice that this list does not contain specific mathematics content. Rather, it is a very abbreviated list of what mathematicians do as they work in the discipline of mathematics. In my opinion, the current mathematics education system can be significantly improved through directly emphasizing the development of math maturity and nurturing it throughout the K–12 mathematics curriculum. We should directly involve students in routinely talking about, doing reflection and metacognition on, and engaging in self-assessment on the progress they are making in improving their level of math maturity.[4]

BUILDING ON THE WORK OF OTHERS

God created the natural numbers. All the rest [of mathematics] is the work of man.
—Leopold Kronecker, German mathematician and logician; 1823–1891

If I have seen further it is by standing on the shoulders of giants.
—Isaac Newton; English mathematician and physicist; 1642–1727

These are two of my favorite mathematics quotations.[5] Mathematicians have been building on the previous work of mathematicians for thousands of years. Our total accumulation of mathematics content knowledge is huge and continues to grow.

Isaac Newton emphasized that our accumulated mathematical knowledge provides a foundation (building blocks) that can be built upon. I find it helpful to view mathematics education as providing students with building blocks and also providing them with lots of problem solving experiences that require thinking and understanding in using the building blocks.

Think about division as a building block. How much does learning the long division paper-and-pencil algorithm contribute to understanding this building block? It takes little time to learn to enter numbers into a calculator and push the division key. It is clear to me that time spent gaining speed and accuracy in division and many other paper-and-pencil calculations can better be spent in gaining understanding and experience.

A computer program that can solve a specific category of mathematics problems, or that makes use of mathematics to help solve such a category of problems, can be thought of as a mathematics building block. Computer Algebra Systems,[6] such as Wolfram Alpha, provide us with a huge number of such building blocks. People who have not studied and learned to understand the underlying mathematics are still able to use such building blocks.

As an aside, think about the software being used by artists who do computer animation in today's film production. The software incorporates a substantial amount of mathematics—content that the users have not formally studied. Their strengths are in the creative use of their artistic talents, and the software is merely a tool that facilitates this artistic creativity.

An organized chunk or collection of valid information is also a type of building block (Moursund, 2014, December). Our schools emphasize students learning to "read across the curriculum"—that is, to read well enough to "read to learn" in each discipline that they study. Computers now provide us with huge libraries that can be accessed online. Thus, students who can "read to learn" and who have access to the Web now have access to building blocks that can provide "just in time" information and instruction.

SOME EDUCATIONAL CHANGES TEACHING MACHINES WILL PRODUCE

I believe we are at an inflection point leading to a time when teaching machines will become a major and eventually a dominant component of K–12 instruction. The teaching machine of the future will be an artificially intelligent, multi-purpose tool designed to provide high-quality individualized instruction. It will be far more than an online delivery vehicle for traditional courses.

As teaching machines gain in intelligence and capabilities, users will want their specific teaching machines to reflect their personal language(s), history, culture, religion, and so on. Various stakeholder groups will want their members to receive an education that is consistent with the values and beliefs of their group. At the same time, we are all "citizens of the world," and thus need an education that prepares us to be responsible, contributing inhabitants of our world.

Development of these teaching machines will require considerable cooperation among a large number of people from many countries. We can get a hint that this is possible by looking at the success of the Web. Its general protocols and functionality provide a relatively uniform service to the whole world. Thinking of the Web as a tool, humans have already created an interactive global tool that is a teacher.

Here is another way in which I believe teaching machines will change education. We are used to dividing a school day into time blocks in which specific topics are taught. However, we know that the problems of the world and the problems that people face in their everyday lives are multidisciplinary.

As more and more instruction is provided by teaching machines, it will become much easier for courses and units of study to contain a combination of domain specific and interdisciplinary content. For example, mathematics is an important tool in many different disciplines. However, many teachers in the non-math disciplines do not feel confident and competent in integrating instruction about mathematics and/or the use of mathematics in the disciplines they teach. Teaching machines will be a major aid to appropriately integrating mathematics instruction and use into the full range of disciplines that their users study.

My feeling is that mathematics educators should be involved in developing the instructional materials in every discipline in which mathematics is a useful tool. Of course, we can make a similar statement about the need for math curriculum development teams to include subject matter experts from non-math disciplines. Although the world needs a large number of highly educated specialists in the various disciplines, it also needs greatly expanded interdisciplinary education for all.

A student's teaching machine will build a profile of its student's capabilities and limitations, interests, likes and dislikes, past and current levels of knowledge and skills, increasing level of physical and cognitive maturity, and so on. The teaching machine will learn from the information it gathers to help improve its student's level of cognitive maturity and capabilities in each discipline the student studies. Formative assessment, and continual feedback based on that assessment, will be routine.

However, this entails a considerable change in the schooling–student interaction. Many students currently view schooling as something that is being inflicted on them, and as a sort of competition in which the goals are to learn to "play the game and to get by" with only a modest amount of effort. This situation does not exist in competitive sports and non-athletic games. So, we know that students are quite capable of engaging in a variety of educational settings in which their individual goals are to work to improve their performance to as high a level as possible.

The challenge in domain-specific and interdisciplinary math education is to make it much more intrinsically motivating to students. Many educational researchers and developers are working on the use of computer technology as an aid to meeting this challenge. For example, you might want to read about serious games.[7] A major industry now focuses on developing games that are specifically designed for use in education (Young et al., 2012).

FUTURE TEACHING MACHINES

The numbered list below provides some of my forecasts regarding future teaching machines. My forecasts are not discipline specific, but all are applicable to mathematics education. Some of these, along with other forecasts, are discussed in two *Information Age Education Newsletters* (Moursund, 2014, September).

1. Future multifunctional teaching machines will handle voice input and output, provide real-time translation of oral and written communications, and be quite portable. When helpful to students, the teaching machine will incorporate wearable technology, and it will include a future version of Google Glass and other data gathering/sensing devices. Thus, for example, a student will be able to look at an object (such as a person, building, tree) and ask the teaching machine to retrieve information about it. Teaching machine responses to such information requests will be contextual, taking into consideration the user, the user's current location, the time of day, and so on.
2. The Web was invented in 1989 and is now by far the world's largest library. *Wikipedia* illustrates how to develop multilingual sources of information. Taken in conjunction with steadily improving connectivity, we have solid evidence that we can provide every student on earth with access to a multilingual comprehensive library that is an integral component of their teaching machine.
3. The first Massive Open Online Courses (MOOCs) were developed in 2011.[8] By making use of data about the performance of all students enrolled in such a course, educational researchers and course developers are rapidly improving MOOCs. My forecast is that eventually such courses will have the combined characteristics of today's HI-ICAL systems and Cognitive Tutors, and in many ways will be more successful than the average classroom teacher. Computer-delivered units of instruction and full courses will be a routine component of teaching and learning, both inside and outside of schools.
4. Currently, the frontiers of computer use in the various disciplines focus on a human accurately specifying a problem to be solved or a

task to be accomplished. Given such a specification or question, the computer or computerized robot takes over the detailed task of figuring out how to solve the problem or accomplish the task. Research and development in IBM's Watson project (Smith, 2014, October 10), as well as by many other research groups, are developing computer systems that can effectively deal with imprecise data and specifications of problems/questions.[9] We as individuals, our educational system, and our legal system all face the major challenge of how far we should trust and act on the judgment and actions (or, proposed actions) of intelligent computer systems. In my opinion, this issue needs to be integrated into each content area of school curricula.

5. Teaching machines will facilitate education to become much more individualized and "hands on." However, individualized does not mean learning in isolation. Students will routinely work in small and large groups, often with a human teacher as facilitator, to solve problems and accomplish tasks that challenge and help advance their current capabilities and interests. Project-based and problem-based learning will become a routine part of a student's life. We see this grouping process now in social networking, online games, and MOOCs. As in current networking, academically oriented groups will routinely consist of students who are widely dispersed in location and age.

FINAL REMARKS

We are at a major turning point in mathematics education. Perhaps more so than in any other discipline, mathematics education lends itself to the concept that *the tool is the teacher*. Mathematics education will become much more individualized through extensive use of teaching machines. Mathematics will be much better integrated throughout the curriculum as students become facile in using computerized tools and as teaching machines integrate this knowledge and these skills across mathematics and all other disciplines.

Throughout most of human history, education has predominantly been a human endeavor. Teaching machines will change the roles that human teachers now play in the current "large classroom group" delivery of instruction. They will provide any place, any time delivery of instruction. This will not obviate the need for teachers to work with individual students and groups of students. Human teachers will provide the human factor in teaching and learning, and will help students to integrate their teaching machine-delivered instruction into their personal and social lives.

I have not attempted to deal with the challenges of restructuring preservice and inservice staff development as we work to implement routine use of teaching machines. In my opinion, this is at least as large a challenge as the one we currently face in attempting to implement CCSSM (Schoenfeld, 2014, September 21). Clearly, the roles of teachers are changing and will change significantly in the future.

I foresee a bright and challenging future for mathematics education!

P.S. TO FINAL REMARKS

The CSMC conference was one of the best conferences I have attended and participated in. Along with many other participants, I found the content to be highly relevant to my interests—and somewhat overwhelming.

In my oral presentation, I mentioned a few ideas from other speakers' presentations that seemed particularly important to me. So much to learn—so little time.

1. We can make math education better by "doing better" in what we have been doing for many years. A number of speakers focused on use of computer technology to do this. Large publishers have committed huge resources to developing computer-assisted instructional materials with this orientation.
2. Some speakers brought me up-to-date on software tools that might be thought of as "super" math manipulatives. New Cabri geometry software is an excellent example and a very powerful aid to learning. See http://www.cabri.com/new-cabri-2-plus.html.
3. The steadily growing capabilities of math problem-solving aids such as CAS received considerable emphasis. What math are students learning and how do we assess this learning when CAS calculators, tablets, and laptops are available to students during summative assessment? Some countries are doing much better than others in exploring these issues.
4. Microcomputer-Based Laboratory tools for use in the sciences are widely available and continue to improve in capability. They provide an excellent example of the tool being the teacher. They also demonstrate how the sciences have developed great tools that students enjoy using, and that bring more life, excitement, and relevance to their study of the sciences. In my opinion, the math curriculum could benefit by making more use of these science tools.

NOTES

1. For more on Alan Kay, see http://iac-pedia.org/Alan_Kay.

2. See http://en.wikipedia.org/wiki/North_American_Aerospace_Defense_Command.
3. See http://iae-pedia.org/Computational_Thinking.
4. See http://iae-pedia.org/Self-assessment_Instruments.
5. See http://iae-pedia.org/Math_Education_Quotations for more quotations.
6. See http://iae-pedia.org/Free_Math_Software#Computer_Algebra_Systems.
7. See http://journal.seriousgamessociety.org/index.php?journal=IJSG.
8. See http://i-a-e.org/iae-blog/entry/supersized-online-courses-moocs.html and Anderson (2014, September 23).
9. See also http://www.research.ibm.com/labs/watson/

REFERENCES

Anderson, N. (2014, September 23). MIT study finds learning gains for students who took free online course. *The Washington Post*. Retrieved from http://www.washingtonpost.com/local/education/mit-report-finds-learning-gains-for-students-who-took-free-online-course/2014/09/23/7ceb34e6-4330-11e4-b47c-f5889e061e5f_story.html

Ask Dr. Math (n.d.). Finding square roots without a calculator. *The Math Forum @ Drexel*. Retrieved from http://mathforum.org/library/drmath/sets/select/dm_sqrt_nocalc.html

Bransford, J., Brown, A., & Cocking, R. (Eds.). (1999). *How people learn: Brain, mind, experience, and school*. Washington, D.C.: National Academy Press.

Card, O.S. (1985). *Ender's game*. New York, NY: Tor Books.

Kuchment, A. (2013, October 21). Teenager creates new flu drugs. *Scientific American*. Retrieved from http://blogs.scientificamerican.com/WSS/post.php?blog=11&post=1445

Moursund, D. (1989). Three 1987 math education scenarios. *IAE-pedia*. Retrieved from http://iae-pedia.org/Three_1987_Math_Education_Scenarios

Moursund, D. (2012). *Using brain/mind science and computers to improve elementary school math education*. Eugene, OR: Information Age Education. Retrieved from http://i-a-e.org/downloads/doc_download/232-using-brain-mind-science-and-computers-to-improve-elementary-school-math-education.html

Moursund, D. (2013). Two brains are better than one. *IAE-pedia*. Retrieved from http://iae-pedia.org/Two_Brains_Are_Better_Than_One.

Moursund, D. (2014a). Brain science. *IAE-pedia*. Retrieved from http://iae-pedia.org/Brain_Science

Moursund, D. (2014b). Improving mathematics education. *IAE-pedia*. Retrieved from http://iae-pedia.org/Improving_Math_Education

Moursund, D. (2014c). Math maturity. *IAE-pedia*. Retrieved from http://iae-pedia.org/Math_Maturity

Moursund, D. (2014d). Problem solving. *IAE-pedia*. Retrieved from http://iae-pedia.org/Problem_Solving

Moursund, D. (2014e). Two brains are better than one. *IAE-pedia*. Retrieved from http://iae-pedia.org/Two_Brains_Are_Better_Than_One

Moursund, D. (2014, September). Education for students' futures: Part 14 and Part 15—The future of teaching machines. *Information Age Education Newsletter.* Retrieved from http://i-a-e.org/newsletters/IAE-Newsletter-2014-146.html and http://i-a-e.org/newsletters/IAE-Newsletter-2014-146.html

Moursund, D. (2014, December). Credibility and validity of information Part 4: The discipline of mathematics. *Information Age Education Newsletter.* Retrieved from http://i-a-e.org/newsletters/IAE-Newsletter-2014-151.html

National Governors Association Center for Best Practices & Council of Chief State School Officers. (2010). *Common Core State Standards for Mathematics.* Washington, DC: Authors.

PACT Center (n.d.). Pittsburgh Advanced Cognitive Tutor Center @ Carnegie Mellon University. Retrieved from http://pact.cs.cmu.edu/index.html

Russell, T. L. (2014). The no significant difference phenomenon. *WCET.* Retrieved from http://www.nosignificantdifference.org/search.asp

Schoenfeld, A. (2014, September 21). Commonsense about the Common Core. The *Berkley Blog.* Retrieved from http://blogs.berkeley.edu/2014/09/21/common-sense-about-the-common-core/#.VPfO5Xdw4LY.facebook

Smith, F. (2014, October 10). Watson could power "tech that teaches teachers." *EdTech.* Retrieved from http://www.edtechmagazine.com/higher/article/2014/10/watson-could-power-tech-teaches-teachers

Sylwester, R., & Moursund, D. (Eds.). (2012). *Creating an appropriate 21st century education.* Eugene, OR: Information Age Education. Retrieved from http://i-a-e.org/downloads/doc_download/243-creating-an-appropriate-21st-century-education.html

Young, M. F., Slota, S., Cutter, A. B., Jalette, G., Mullin, G., Lai, B., . . . Yukhymenko, M. (2012). Our princess is in another castle: A review of trends in serious gaming for education. *Review of Educational Research, 82*(1), 61–89.

CHAPTER 19

DEEPLY DIGITAL STEM LEARNING

Chad Dorsey

I am honored to help close such a great conference. I'm especially pleased to be able to follow such a great group of presenters, including such an inspiring set of looks at the classroom—having been a teacher and done professional development for a number of years, I fully appreciate the importance of having a view from the classroom.

I was asked to talk about something a bit different to draw some distinctions and parallels; I'm going to talk about Deeply Digital STEM Learning. There has been much discussion at this conference about what makes digital resources distinctive. I have been excited to see this, as it really brings full circle some of the discussion we started about four years ago at the Concord Consortium and elsewhere. It was particularly fun to go back through a couple of talks from back then and be able to play out some of the messages that have been resonating these past couple of days.

One of those messages has been about the power of tools. I'm going to show some STEM examples today and hopefully show the links they have to important mathematical ideas. I'll also show some fun toys and cool software in the process. One of them is what I have in my hand, an infrared

(IR) camera. These used to cost about $10,000 not more than 6–7 years ago. Then they dropped to less than $1000 and became handheld cameras in the last five years—you can rent one at Home Depot now to look at your home's insulation. Now they've become this, a $300 attachment to an iPhone, and even more impressive than the original. It's amazing. And they can see tiny differences in temperature. Pressing my hand on the desk for ten seconds makes the handprint visible—you'll be able to see that for several minutes.

This is a technology that does what truly transformative technologies do—enable you to take a fresh look at the world around you. I want to start by showing you something you've seen every day. This regular glass of water is at room temperature. Viewing it through the IR camera, you can see that the water is blue, meaning it's colder than the surroundings (because it's evaporating—as happens to your body when you get out of the pool). Now I'm covering one side with a piece of paper—note what happens. The covered side becomes red—that means that it's hotter than the surroundings, and hotter than the water itself!

How can this be? Energy isn't supposed to flow in that direction!

But before I talk more about this example of deeply digital learning, I have to tell a brief story about my childhood. It's nice that the conference was held in Chicago, because I'm at home in the Midwest; I grew up in Minnesota with the sweet corn, hot dish potlucks, and butter sculptures of beauty queens. Now, anyone who's heard Garrison Keillor knows Minnesotans can talk. So every family visit I can remember set aside at least 30–60 minutes on each end for discussion—about navigation. "Did you take 35? 218 is faster. Really? Well, the cutoff on 77 is actually best..." You get the idea.

Why am I talking about this? Because it's a really hard problem that used to seem impossible—remember the traveling salesman problem? Bubble-sort algorithms? But today, with GPS, it's almost completely gone. Get in. Press Home. Drive. Now, it's a little-known fact that this problem had been addressed before. I have brought some technology to demonstrate—this paper map.

Remember paper maps? They are obviously very advanced technology: expandable full-screen mode, extended zoom capabilities, even a built-in search feature...

The paper map is good technology, but it has been almost completely supplanted in no time at all. And the role of once-static maps has fundamentally changed, too. Today, our maps dynamically show us where we are right now—that's a subtle, but immense shift. We can focus entirely on our current location and the context that surrounds us, and adjust accordingly. The little blue dot showing our current location represents a huge transformation.

A similar level of transformation is poised and beginning in learning. One thing that has been discussed a bit during this conference is how we have a duty to be prepared for the inevitability of this change. There's a quote from Chris Dixon (2011) I love: "Predicting the future of the Internet is easy: Anything it hasn't yet dramatically transformed, it will." Anything the Internet hasn't changed yet? Just wait. We've seen this happen over and over.

We talked about this at this conference. And we've seen it in the world around us. The music industry tipped in a few years. Entertainment and news aren't taking much longer. Education is only a matter of time. Every single one of you will be part of the education revolution—the question is just whether you will be driving it or following it. And as has been exhorted, you need to be prepared for the idea of just how quickly and radically these changes will come—I'd like to take a minute to appreciate just how quick and radical a change we need to prepare for. Remember that when we're talking about the Internet, it puts dog years to shame.

So I'd like to do a little experiment, to think ahead by thinking back. Consider the students who are freshmen in high school today. What will the world be like for them when they graduate? To answer that, the best thing we can do is to look back four years. When we do, we realize that the iPad *did not exist* four years ago—tablets and the basis for so much of the stuff we've been discussing at this conference weren't even *introduced*. We should expect a similar innovation that we can't anticipate by the time they graduate. Looking back a step further, in the time it takes today's middle school students to graduate, we can expect innovations on the order of the introduction of Twitter, the introduction of Gmail, and the introduction of the iPhone itself.

Let's continue—in the time it takes today's third-graders to graduate, if we look back in time, we can expect that we'll see new innovations on the order of Google Docs, YouTube, Google Maps and Google Earth, and Facebook. All of those were introduced to the world in an equivalent time backward. And in the time it takes today's preschoolers to graduate from high school? In that time looking back, Wikipedia was founded, the iPod was introduced to the world, accurate GPS signals became available for the first time, and Google was first funded. If we think one more step back, in the time it takes children born in six years to graduate, we can expect innovations on the order of the first DVD, the first Web browser, the modern cell phone network, and the first Web server. So the equivalent of another World Wide Web scale innovation is what we can expect for the students who will be coming into our kindergartens in barely a decade. And we can't even remember today what life was like before these innovations.

We have to be prepared for textbooks to transform as radically as maps have, and for changes we can't predict at all. So, what if we can't predict what will come, and it will come so rapidly we can't avoid it? What should we be doing?

This is a question we think about a lot at the Concord Consortium. We're a non-profit organization in Concord, MA and now in the Bay Area as well. We are a group of innovative scientists, educators, curriculum developers, and software developers who work in creative collaboration to transform STEM education through technology. We research and develop tools that make the invisible visible and explorable and help bring out the inner scientist in everyone. We've spent some time thinking about what today's students need and about the elements of deeply digital learning.

In our initial work, we defined something like seven different aspects of deeply digital learning, and I think we probably missed a couple, but many have come out in the talks here. The core of it is—and this is the key—to be asking the question: Are we using technology to its fullest potential for learning? (And along with that, what is technology adding at all points of the learning process?) Answer that question, and we meet the—inherently shifting—notion of deeply digital learning. So what should we be doing so we can sleep well at night knowing we're doing right by our children?

Fortunately in science we have a few benefits we can count on. Several of them are connected to the Next Generation Science Standards (NGSS). One benefit is that they are coming second, behind the Common Core State Standards (National Governors Association Center for Best Practices & Council of Chief State School Officers, 2010), so that folks who think about teaching and learning in science can sit back a bit and watch what happens with the Common Core State Standards rollout and benefit by knowing what mistakes not to make. The second benefit is that they build upon a great foundation. The National Research Council Framework for K–12 Science Education is a great document, thought out well by an excellent group of people. One of the intriguing things about this document is that it does not use the word "inquiry," though many would expect it to. The reason is that if you ask 7 people what inquiry means, you'll get 9 definitions.

Because the concept of inquiry has been diluted so significantly, the group developing the Frameworks looked back at things and instead defined some different elements. These include the Practices of Science and Engineering, which are the things that students should be engaging in as they undertake scientific endeavors both in the classroom and the real world. These have analogs in the Common Core State Standards as well, so they are somewhat familiar ground. These practices are shown in Table 19.1.

These practices exist as a core element of the NGSS, which also includes a set of crosscutting concepts, such as Energy, Structure and Function, and Systems. All of these elements are taught together—you can't teach a

TABLE 19.1   The NGSS Scientific and Engineering Practices

- Asking questions/defining problems
- Developing and using models
- Planning and carrying out investigations
- Analyzing and interpreting data
- Using mathematics and computational thinking
- Constructing explanations/designing solutions
- Engaging in argument from evidence
- Obtaining, evaluating, and communicating information

practice of science without using the content it refers to, and you can't teach a crosscutting concept in isolation, either. Figure 19.1 shows a schematic of the NGSS Practices, Disciplinary Core Ideas, and Crosscutting Concepts.

By taking one or more elements from each layer, we obtain triads that can guide teaching and selection of resources. We may want to use models with life science ideas to help students understand and appreciate scale and structure/function, for example. I'd like to look at a few of the practices a bit more deeply, because these are in some ways where technology really starts to make a difference.

The first of these, Asking Questions, is one we engaged in earlier, when we discussed the images from the IR camera. What ideas did you come up with when you thought about this? Why did the half of the paper covering the water get warmer than the surroundings? In fact, we were confused by this as well and we had to do some extensive experiments—Charles Xie, the author of the Molecular Workbench, who began using these IR cameras in our labs, spent weeks doing experiments to confirm what we observed.

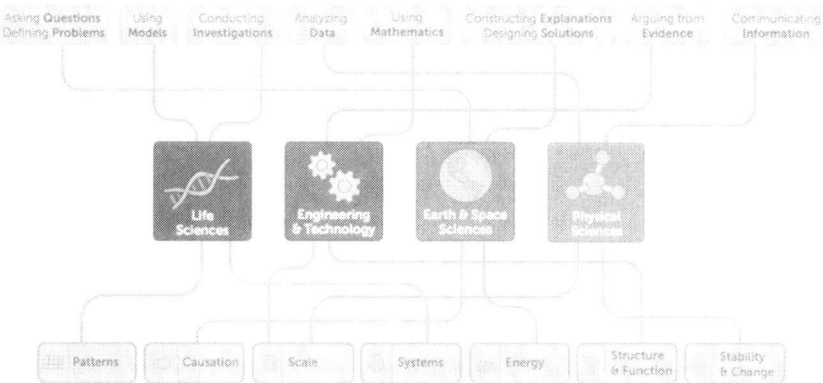

Figure 19.1   The NGSS practices, disciplinary core idea areas, and crosscutting concepts.

What's actually happening when water evaporates? The highest energy water molecules near the surface are the ones with enough energy to escape the surface of the water—it turns out that those are landing on the paper-covered half, and giving up their thermal energy to the paper. The amazing thing—we did the calculation—is that you're actually seeing a layer of water only 10 molecules thick on average!

This is happening in any glass of water you drink every day—a clear example of technology making the invisible world around you visible in new ways. And the questions you asked about it were immediate and authentic. Asking questions is something we can all do, but technology enables us to ask questions in new ways, and about concepts and phenomena that would be otherwise inaccessible.

Another example of a quite relevant and important NGSS Practice is Developing and Using Models. This is a key skill for students, and also has some relevance over time, as students begin to see models underlying many, many systems and examples. One of the key underlying mechanisms driving everything around us is the invisible world of molecular dynamics and its ever-present phenomena. Now, it's hard to lecture about atoms and molecules. And you can't really do a lab experiment with them. Fortunately, we've addressed that issue with our Molecular Workbench software (mw.concord.org/nextgen). By taking research-grade algorithms for molecular dynamics and placing a pedagogical layer on top, we provide students a laboratory in which they can do experiments on the molecular world. Figure 19.2 shows an example of an interactive environment in which students can shake up oil and water molecules and see them recombine—understanding in a

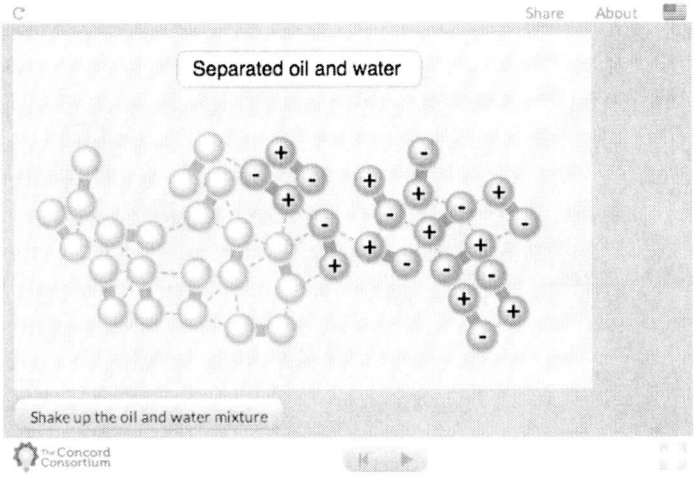

Figure 19.2 A Molecular Workbench interactive.

straightforward manner how intermolecular forces drive a phenomenon such as the separation of polar and non-polar liquids. We have hundreds of these models and scores of activities in which students can explore the molecular world that underlies all of physics, chemistry, and biology.

Another important practice of scientists is argumentation supported by evidence. Scientists find themselves making arguments throughout all their activity, both formally and informally, supporting these arguments at all stages with evidence they have derived from experiments. We want students to be doing the same thing as they learn science—to be conducting science as scientists do. This is more complex in some domains than in others, however. For example, genetics is a field that is revolutionizing modern society—there could hardly be a more important area in the next decades than genetics—but it is very hard to access for teaching and learning. You can't see genetics. You can't do experiments with genetics, unless you count fruit flies in AP Biology, and even then you end up only getting a couple generations. Furthermore, genetics is very complex because of the multiple scales involved. Changes that occur on the molecular level cause visible changes in whole organisms, and the whole process is mediated by random cellular-level mechanisms.

So we've created a modeling environment that permits students to explore these ideas in a virtual laboratory, and done so in a fun way by allowing students to breed dragons. In Geniverse (geniverse.concord.org), students can manipulate alleles and see instant changes in the organism, as shown in Figure 19.3. They can examine chromosomes as they sort into

Figure 19.3 Students manipulate dragon genes and observe the effects in Geniverse.

eggs and sperm cells and can cause changes and see their effects. And, they can examine the effects of their experiments and make claims supported by evidence to describe how they think various traits are inherited. These claims go into a class journal, and students bring them back the next day to the classroom to discuss and argue for their points of view. When they find conflicting points of view, they return to the virtual laboratory and do experiments to test their ideas, just as practicing scientists do. By providing space for students to perform these experiments, we enable them to get closer to the phenomena themselves.

Another important practice of science and engineering is designing solutions. Especially in engineering, students need to be able to design original solutions to a problem and see how they function. Technology can enable this in many ways. To help students do this, we have created Energy3D (energy.concord.org), an environment that permits students to design solar heated houses and test them out. As seen in Figure 19.4, within minutes students of all ages can use Energy3D to design a house and orient it for maximum use of incoming solar energy. They can test out the solar energy influx at different latitudes, and can even generate a set of two-dimensional plans they can cut out and assemble to make a real-life model of the house. With this model, students can use real probes and sensors along with IR cameras to test and refine their designs. This mixture of virtual tools and

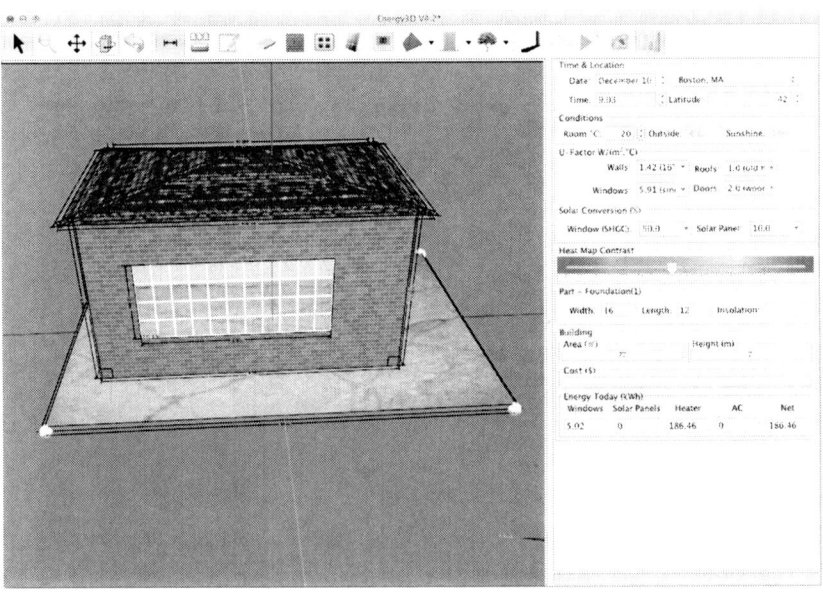

Figure 19.4 Students design houses and test their energy efficiency with Energy3D software.

hands-on experimentation takes the best of both worlds and uses technology to enhance students' ability to design solutions as fully as possible.

One of the most central aspects of all of science is the process of experimentation. In the NGSS practices, this is broken down to a series of different stages, one of which is called Planning and Carrying Out Investigations. Another is called Analyzing Data. Clearly, both of these are essential parts of the process of investigating the world around us in the way that scientists do. Technology can be a large help in this process. Hands-on experimentation is significantly aided by the addition of technology-enhanced data collection.

Some of you may recognize Figure 19.5 as the auto-focus sensor on Polaroid cameras. These were provided as part of an engineering kit from the Polaroid company over thirty years ago, and our team got them and started to experiment with them, hooking them up to an Apple II computer and cycling the pulses, then graphing the results. This generated the first motion graphing probe, which later became a commercial success for many different companies, and was part of a revolution in scientific data collection. Bob Tinker, our founder, traveled around the country demonstrating data collection via probeware and showing how an experiment that usually took two class periods could be performed in 30 seconds using a temperature sensor and an early computer. Now educational probes and sensors are developed by 5 major companies worldwide and over 1 million students annually use them for data collection.

The collection of data from probes and sensors fundamentally changes the process of experimentation. The emphasis can shift from the mechanics of making a graph to the process of interpreting it. The immediacy of feedback derived from the action of this data collection brings the learner closer to the activity being measured, and helps solidify understanding. We have built on this for decades, and are now doing work that brings this process to a new stage as well.

When students take scientific data and examine it, another fundamental leap is involved. Scientific data typically comes from a set of individual

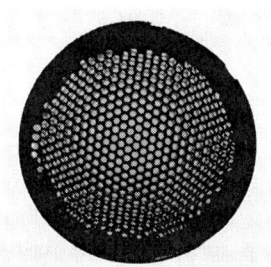

Figure 19.5 A perforated disk.

experimental runs. Summary variables from the different runs need to be aggregated and analyzed. This creates the need to understand hierarchical data and understand and manipulate the aggregate, summary data. In this process, students must navigate mentally in a parameter space, identifying places where new data are needed, and understanding that individual data points are proxies for full data sets themselves. This "parameter-space reasoning" is a critical skill for understanding and manipulating scientific data sets. However, it involves the collection, aggregation, and analysis of significantly more data—a process that is not afforded by current technology setups. Our InquirySpace technology is working to ease this while also providing access to multiple representations and ways of thinking about important phenomena.

InquirySpace unites three important technologies: probes and sensors, models and simulations, and a data exploration environment. The last of these is likely familiar to many here—those who are familiar with Fathom or Tinkerplots likely know Bill Finzer, the creator of Fathom at KCP Technologies. Bill Finzer is now a Concord Consortium employee, and Fathom is being distributed (in a free, time-capped version, for the time being) by the Concord Consortium as well. Bill is working with a team to develop a new environment that is very much the online successor of Fathom and TinkerPlots, our Common Online Data Analysis Platform (CODAP). CODAP is being developed as web-based, open source software through a three-year National Science Foundation grant that began this year. It has already been integrated into our InquirySpace project in beta form. By using CODAP, students are able to explore data openly and freely as is the case with Fathom and TinkerPlots. Moreover, now students are able to take in data from multiple sources, including probes and sensors and models and simulations.

Using InquirySpace (inquiryspace.concord.org), students can gather data from a mass and spring and import the data into CODAP, where they can analyze it further, as shown in Figure 19.6. We don't specify the axes of the graph, but instead let the user determine them by dragging variables to the axes. This open exploration affords many new discoveries by students. We are developing the CODAP project by working together with multiple other projects. For example, EDC's OceanTracks project is providing data about GPS-tagged marine mammals that we are using to help develop affordances for exploring time- and location-tagged GIS data. The possibilities are truly enormous when students are able to examine data of all types. And, we think this is vitally important, as the goal is to get more students using more data in more places. We know that data science has arrived as an area of major importance for our society. We think that it's essential that data science education follow suit. As it does, we're preparing for data

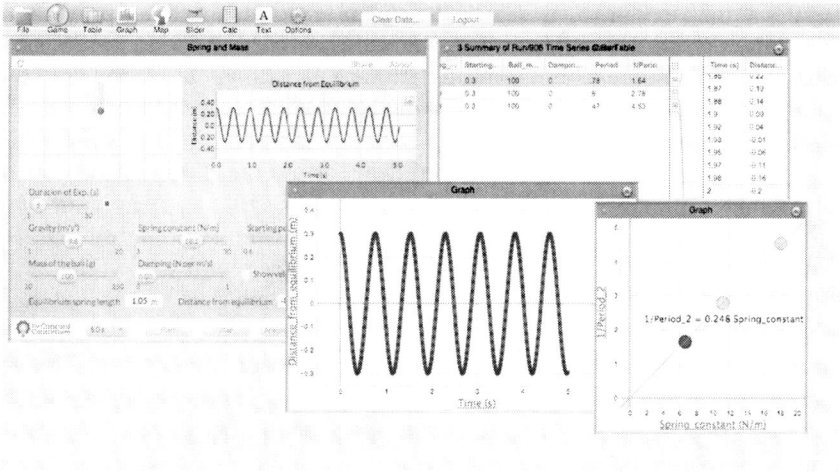

Figure 19.6 Data from a mass/spring simulation analyzed using InquirySpace software.

science education research, which we'll need to understand what is and what isn't working in the process.

As we do this, we are able to take advantage of one thing that hasn't been talked about enough at this conference. A critical aspect of deeply digital materials is the fact that we can know so much about how users interact with them, and that information can serve as essential data about their learning. Learning analytics and data mining are extremely promising new frontiers in educational technology. In the examples I just demonstrated in CODAP, we were logging behind the scenes every action I took at fine-grained detail. We can then know when students create graphs, how they interact with simulations, what data they look at and what data they don't look at, and many other things. By analyzing these data about student interactions, we can learn a huge amount about what students are doing and learning. For example, we can look at students' interactions with a game embedded in CODAP and parse the learners into seven distinct groups just from the data. We can see that the exercise didn't challenge some students, and can see that another group wasn't able to approach it properly.

In other places, we are able to see fascinating patterns that couldn't be seen otherwise—detailed log data from students using Energy3D to design houses allows us to see whether students are approaching the task iteratively or with introspect into their design process. We're able to see gender differences in the design process from looking at the data alone, in many ways starting to tap into a holy grail of engineering design. These are far frontiers indeed, and we see this as a ten-year problem at least, but the importance is clear—if we can start to understand whether students are engaging in inquiry or the engineering design process productively, we can start to assess elements of learning that have remained elusive for decades, and possibly begin to find ways to avoid the trials and tribulations brought by today's large-scale assessments.

We do much more at the Concord Consortium that I don't have time to show here. We have scores of curriculum activities that we deliver through an online platform, embedding models and simulations and permitting students to learn in new, richer ways. These platforms report student progress to teachers and permit the teachers to customize materials as well, an aspect that has been emphasized frequently at this conference. If we bring all of these elements together, only then can we aim for the promise of what we're aiming for, learning that truly uses technology to its fullest potential—deeply digital learning. If we all work together and get to that stage, there's no telling where we can go.

Thanks to the National Science Foundation for funding many of our projects and to Google for funding the translation of our Molecular Workbench software to HTML5. We encourage everyone to follow our newest developments by following us on Twitter at @ConcordDotOrg and @chaddorsey, and by liking us on Facebook at facebook.com/concordconsortium.

REFERENCES

Dixon, C. (2011). *Predicting the future of the Internet is easy: Anything it hasn't yet dramatically transformed, it will.* Retrieved from http://cdixon.org/2011/01/13/predicting-the-future-of-the-internet-is-easy-anything-it-hasnt-yet-dramatically-transformed-it-will/

National Governors Association Center for Best Practices & Council of Chief State School Officers. (2010). *Common Core State Standards for Mathematics.* Washington, DC: Authors.

CHAPTER 20

CLOSING REMARKS

Zalman Usiskin

During WWII, media began to be used by both sides for "education." Specifically, radio and movies were used for propaganda. Soon afterwards some of the best minds were used to create movies. In 1959, Walt Disney's cartoon film *Donald in Mathemagic Land* was released. In 1966, the film *Let Us Teach Guessing*, with George Polya, was made available by the Mathematical Association of America. But television became the primary media vehicle for overt attempts at education. There were 44 educational television stations in the United States in 1960, but 175 by 1969. Gifted teachers such as Lola May, who was at that time the speaker at National Council of Teachers of Mathematics meetings who would draw the most attendees, were enlisted to create full television courses in mathematics.

Also around this time, the impact of computers was felt, not in computer-assisted instruction, which was to come later, but in programmed learning. Full courses were written by IBM and others whose intent was to enable students to proceed, bit by bit, through a course by responding to an item and moving on to a next question if the response was correct or having some corrective action if not. This allowed students to move at their own pace and, in theory, gave all students the advantage of an education from the finest of materials.

These technological advances—movies, television, and computer ideas—all were about the *delivery* of curriculum, not the content, which was little changed. But then hand-held calculators appeared in the very early 1970s. Quite quickly books at the high school level appeared in which tables of square roots, tables of trigonometric functions, and logarithms for computation were not present. But how slowly the task has been to rid the elementary school curriculum of obsolete algorithms. Just last week it was reported that the sample 6th grade Partnership for Assessment of Readiness for College and Careers (PARCC) test contains a division problem to be done by hand in which the divisor is a 3-digit decimal. This is not an anomaly. Multiplication and division of decimals with limited numbers of decimal places are 5th grade Common Core State Standards for Mathematics (CCSSM); the number of decimal places is unlimited in 6th grade.

In the middle 1980s, graphing calculators appeared. Their ability to graph functions is indisputable, high school teachers have tended to embrace them, and they are allowed on our college entrance and AP exams, but the reaction of many college-level mathematicians is as if they are the plague. Don't they realize that generations of college calculus students have memorized formulas for derivatives and integrals without knowing much else about them? Geometry drawing programs also appeared around this time, but so far they have had little if any effect on the curriculum.

In the middle 1990s, hand-held computer algebra systems appeared. These enable the user to see patterns in working with expressions and equations that are particularly useful for the crippled algebra student. In some places in the world, including—as we learned at this conference—in Denmark, these CAS calculators have been welcomed; in the United States only a couple of curricula dare to take advantage of them. We at the University of Chicago School Mathematics Project are proud to be one of those curricula.

In general, these more recent advances have been controversial because they are about changing how students in school *do* mathematics, not about how mathematics is delivered to them. Yet should these students ever use mathematics outside of school, they will find that most often they are expected to utilize the capabilities of computers, not shun them. It is a mismatch that cannot last forever.

Now, in the past decade we have seen the development of tablets and other computers that have the capabilities both to deliver curriculum and do mathematics. This is a powerful combination, and it has led many people to predict that textbooks will be obsolete in a few years, that the classroom will change dramatically, and that, along with this, so will the role of the teacher. However, all these are aspects of the delivery system, and the delivery system has proved to be exceedingly difficult to change. Mathematics in most classrooms is remarkably similar to what it was generations ago.

As many of the speakers noted, digital technology has been touted as helping to adapt the curriculum to the individual. However, as I think Janine Remillard pointed out (see Chapter 13), research on individualized instruction has consistently shown that it is more difficult to learn alone than with a group of others. If the level of student performance remains the barometer by which we judge the effectiveness of new developments, then adaptations of digital technology to the individual may prove to be disappointing.

Yet simultaneous with these developments has been the extraordinarily fast rise of social media, which has the power to turn a classroom of 25 students into a network of thousands of students communicating with each other about mathematical concepts. It is hard to realize that we are still in the infancy of social media. Of the four largest social media sites, Linkedin was launched 11 years ago, Facebook 10 years ago, Twitter 8 years ago, and the development of Pinterest began less than 5 years ago. It is not surprising that we have not yet seen large-scale curricular materials that take full advantage of social media.

Brian Lemmen's and Loretta Asay's presentations (see Chapters 10 and 8, respectively) remind us that learning is as much a social experience as a cognitive experience. We learn our native language by interacting with our parents and others from infancy on up. Classrooms in school stimulate a student to learn in order to keep up with classmates and for self-esteem. Children in a class often remark that they understand an explanation given to them by a classmate better than a teacher's explanation, even when the classmate's explanation is not precisely correct. We take our best students and put them in special classes to stimulate them further. Nations create experiences for their Olympiad teams, realizing that students often learn as much or more from each other than from their teachers. It may be that only when we integrate the social media capabilities of technology into our changes in the delivery system and how we do mathematics will we obtain growth in student performance in mathematics.

It is clear that we are only in the early stages of the delivery and enactment of digital curricula. Every speaker at this conference has spoken of issues raised by dealing with digital curricula. I would like to comment on these.

WHO IS THE EXPERT?

Most of us who have taken on the task of trying to create full curricula are appalled at those who think that a typical teacher, even a typical *good* teacher, possesses the ability to create a coherent curriculum by amassing materials from a variety of sources. A well-structured curriculum has an overt table of contents but it also has subtlety after subtlety that has determined why particular problems have been chosen, and how they have been sequenced, why the lessons are ordered as they have been, and why certain units precede others. A good course has a language that has been introduced early and is consistently used throughout. It has, in the patois of the CCSSM, focus and coherence.

There are many possible good sequences of mathematics experiences for learners at every grade level, but creating any one of them takes expertise. The typical teacher has neither the time nor the breadth of knowledge to make good choices in a consistent fashion. The teacher may know her students best, but may not know what is best for the students. Trying to make teachers into curriculum experts is, in my opinion, an unwise strategy.

SEQUENCING THE CURRICULUM

One characteristic about the field of mathematics that is not true of most other fields is that mathematics is logically sequenced. We define terms using other terms that have been previously defined or are understood. We justify a theorem with definitions, postulates, and previously proved theorems, so the theorem must follow those previously discussed ideas. We know that the algorithm we call "long division" works because we can justify this algorithm based on properties of integers. We can perform this algorithm if we can perform some of its subroutines—estimating how many times a one-digit number goes into a two-digit number, multiplying by a one-digit number, and subtracting multi-digit numbers. Not only are the propositions of mathematics sequenced, so are the algorithms.

At the elementary school level, the sequence of *algorithms* has generally determined the sequence of topics. Gradually, as a student moves up the grades, the sequence of topics becomes more and more determined by the *logical* sequence of mathematics.

Even the simplest calculator technology throws a monkey wrench into this fundamental aspect of mathematics. With a calculator, you have all sorts of algorithms at your disposal and you need not know any of the subroutines that were prerequisite with paper and pencil. The availabiliy of digital delivery systems likewise throws a monkey wrench into the logical order. If I want to know the properties of parallelograms, I do not have to deduce them—I can just look online. If I want to know how to solve a system of 3 linear equations in 3 unknowns, I just go online to find out. Not only may I have violated the logical order, I probably don't even realize that there is a logical order.

ELEMENTARY VS. SECONDARY

All of this leads me to believe that if digital curricula created by teachers out of the plethora of materials out there become commonplace, there will be a backlash to reassert the logical priorities of mathematical topics, and the desire to have a more-or-less fixed sequence will be stronger as one goes up the grades. In this regard and perhaps in others, the use of digital delivery systems that affects the sequence may need to have a rather different look in secondary schools than it has in primary and elementary schools.

Those are just a few of the ideas and issues stimulated in me by the conversations and papers emerging from the conference. I hope that you too have found these three days to be stimulating and worth the time and effort you have spent here.

Allow me to finish with some more well-deserved thanks. First, again to the program committee and the CSMC leadership group. Second, to the local arrangements committee and to the many other people associated with the Center for the Study of Mathematics Curriculum who assisted before and during this conference.

And finally, thanks to you, for gracing us with your presence, your questions and reactions, and your tweets. Without you there would be no conference. We wish you a pleasant rest of your stay in Chicago and a safe trip home or wherever you are going from here. Farewell and fare well!

APPENDIX A

CONFERENCE PROGRAM

**CSMC Third International Curriculum Conference
Mathematics Curriculum Development, Delivery,
And Enactment In A Digital World**

November 7–9, 2014

**Ida Noyes Hall
The University of Chicago
Chicago, Illinois USA**

Friday, November 7, 2014

8:00 a.m.–5:00 p.m.	**Registration (Lobby)**
8:00–9:00 a.m.	**Continental Breakfast** (Cloister Club)
9:00 a.m.	**Setting the Stage: Demonstrations of Digital Materials** (Palevsky Cinema) Presider: Robert Reys, University of Missouri
9:00–9:20 a.m.	Hee-chan Lew, Korean National University of Education *Quadratic Curves in a Digital Textbook*
9:40–9:50 a.m.	Break
9:50–10:10 a.m.	Jere Confrey, North Carolina State University *Math Projects*

10:10–10:30 a.m.	Nathalie Sinclair, Simon Fraser University, Canada *TouchCounts: Using Fingers, Eyes and Ears to Learn to Count, Add and Subtract*
10:30 a.m.	**Refreshments** (Cloister Club)
10:50 a.m.	**Setting the Stage: Demonstrations of Digital Materials** (continued) (Palevsky Cinema) Presider: Steven Ziebarth, Western Michigan University
10:50–11:10 a.m.	• Michal Yerushalmy, University of Haifa, Israel *Designer Views on the Design of the VisualMath e-Textbook*
11:10–11:30 a.m.	• Kaye Stacey, University of Melbourne, Australia *Specific Mathematics Assessments that Reveal Thinking (Smart Tests)*
11:30–11:40 a.m.	Break
11:40–12:00 p.m.	• Jean-Marie Laborde, Cabri and University of Grenoble, France *The New Cabri*
12:00–12:20 p.m.	• Angela Crouse and Chad Idol, McGraw-Hill Education *Engaging, Efficient, Effective, and Easy to Use Digital Tools From McGraw-Hill Education*
12:20 p.m.	**Lunch** (Cloister Club)
1:15 p.m.	**Welcoming Remarks** (Palevsky Cinema) Zalman Usiskin, University of Chicago Barbara Reys, University of Missouri Martin Gartzman, University of Chicago
1:30 p.m.	**Plenary Session I** (Palevsky Cinema) Presider: Christian Hirsch, Western Michigan Univ. Speakers: • Mogens Niss, Röskilde University, Denmark *Mathematics Standards and Curricula—Different Notions, Different Meanings, Different Roles in Different Parts of the World Under the Influence of Digital Affordances* • Kenneth Ruthven, University of Cambridge, England *The Re-Sourcing Movement in Mathematics Teaching: Some European Initiatives*

3:00 p.m.	**Break**

3:15 p.m.	**Plenary Session II** (Palevsky Cinema)
	Presider: Megan Bates, University of Chicago
	Speakers:
	• Kaye Stacey, University of Melbourne, Australia
	  Mathematics Curriculum, Assessment, and Teaching for Living in the Digital World
	• Michal Yerushalmy, University of Haifa, Israel
	  Inquiry Curriculum, Textbooks & Assessment: Technological Changes That Challenge the Representation of School Mathematics

4:45–5:30 p.m.	**Reception** (Lounge/Library)
5:30–7:00 p.m.	**Dinner** (Cloister Club)

Saturday, November 8, 2014

8:00 a.m.–4:00 p.m.	**Registration** (Lobby)
8:00 a.m.	**Continental Breakfast** (3rd floor Theatre)

8:45 a.m.	**Plenary Session III** (Palevsky Cinema)
	Presider: Amanda Thomas, Pennsylvania State University–Harrisburg
	Speakers:
	• Phil Daro, SERP (Strategic Education Research Partnership) and Pearson
	  Pads in the Classroom: Thinking in Public
	• Jere Confrey, North Carolina State University
	  Building Digital Curriculum for Students' Productive Struggle Using Challenges, Tools and Interactivity

10:15 a.m.	**Refreshments** (3rd floor Theatre)

10:30 a.m.	**Parallel Participant Interaction Sessions With Speakers**
	• Session 1 (Palevsky Cinema)
	    Presider: Martin Gartzman, University of Chicago
	    Discussion with:
	    – Phil Daro, SERP (Strategic Education Research Partnership) and Pearson
	    – Jere Confrey, North Carolina State University
	• Session 2 (2nd floor East Lounge)
	    Presider: Kathryn Chval, University of Missouri–Columbia
	    Discussion with:

- Mogens Niss, Röskilde University, Denmark
- Kenneth Ruthven, University of Cambridge, England
- Session 3 (2nd floor West Lounge)
  Presider: Betty Phillips, Michigan State University
  Discussion with:
  - Kaye Stacey, University of Melbourne, Australia
  - Michal Yerushalmy, University of Haifa, Israel

11:30 a.m.	**Lunch** (3rd floor Theatre and 1st floor Library/Lounge)
12:30 p.m.	**Panel: Research On Digital Curricula** (Palevsky Cinema)

- Jeffrey Choppin, University of Rochester (Moderator)
  A Typology for Analyzing Digital Curricula in Mathematics Education
- A.J. Edson, Michigan State University
  A Design Experiment of a Deeply Digital Instructional Unit and Its Impact in High School Classrooms
- Janine Remillard, University of Pennsylvania
  Keeping an Eye on the Teacher in the Digital Curriculum Race
- Nathalie Sinclair, Simon Fraser University, Canada
  New Starting Points for Number Sense Using TouchCounts

2:00 p.m.	**Break**
2:15 p.m.	**Plenary Session IV** (Palevsky Cinema)

Presider: Andy Isaacs, University of Chicago
Speakers:
- Jean-Marie Laborde, Cabri and University of Grenoble, France
  Technology Enhanced Teaching/Learning at a New Level With Dynamic Mathematics as Implemented in the New Cabri
- Hee-chan Lew, Korean National University of Education
  Developing and Applying "Smart" Mathematics Textbooks: Issues and Challenges

3:45 p.m.	**Refreshments** (3rd floor Theatre)

4:00–5:00 p.m.	**Parallel Participant Interaction Sessions With Speakers** • Session 1 (Palevsky Cinema) Presider: Barbara Reys, University of Missouri Discussion with: – Jeffrey Choppin, University of Rochester – Janine Remillard, University of Pennsylvania • Session 2 (2nd floor East Lounge) Presider: Jon Davis, Western Michigan University Discussion with: – A.J. Edson, Michigan State University – Nathalie Sinclair, Simon Fraser University, Canada • Session 3 (2nd floor West Lounge) Presider: Jack Smith, Michigan State University Discussion with: – Jean-Marie Laborde, Cabri and University of Grenoble, France – Hee-chan Lew, Korean National University of Education

Sunday, November 9, 2014

8:00 a.m.–12:15 p.m.	**Registration** (Lobby)
8:00 a.m.	**Continental Breakfast** (Cloister Club)
8:45 a.m.	**Panel: Digital Curricula In Practice** (Palevsky Cinema) • Valerie Mills, Oakland Schools, Michigan (Moderator) *Connections and Distinctions among Today's Digital Innovations and Yesterday's Innovative Curricula* • Loretta Asay, Clarke County Schools, Nevada *Technology to Support Mathematics Instruction: Examples from the Real World* • Josephus (John) Johnson, Battle High School, Columbia, Missouri *We Opened a New School and Gave Everyone iPads . . . We Are so Much Smarter Now* • Brian Lemmen, Holland Christian High School, Michigan *Deeply Digital Curriculum for Deeply Digital Students*
10:15 a.m.	**Break With Refreshments** (Cloister Club)

10:30 a.m.	**Plenary Session V** (Palevsky Cinema)
	Presider: Jeffrey Shih, University of Nevada–Las Vegas
	Speakers:
	• Chad Dorsey, Concord Consortium, Massachusetts
	  Deeply Digital STEM Learning
	• David Moursund, University of Oregon
	  Mathematics Education Is at a Major Turning Point

12:00–12:15 p.m.	**Closing Remarks** (Palevsky Cinema) Zalman Usiskin, University of Chicago
12:15 p.m.	**Lunch** (eat in or to go) (Cloister Club)

APPENDIX B

SPEAKER BIOGRAPHIES

Loretta Asay serves as the Instructional Technology and Innovative Projects Coordinator in the K–12 Mathematics and Instructional Technology Department of Clark County School District, Las Vegas, Nevada. Her responsibility is to increase the appropriate use of technology in classrooms to support student learning. This involves managing several projects, such as the "e3 1:1 Project" which provides devices for over 15,000 students, 24/7, and including a robust professional development program. Her staff works closely with all departments providing professional development and grants for schools, helping them model appropriate technology use in all content areas. Dr. Asay earned a PhD from the University of Nevada, Las Vegas, in Educational Psychology. Research interests include learning theory and the importance of background knowledge, especially looking at how technological tools can support that.

Email: asayl@interact.ccsd.net

Jeffrey Choppin directs the mathematics education program at the Warner Graduate School of Education at the University of Rochester. His methods courses challenge students' conceptions of mathematics and the teaching of mathematics while exploring the influence of societal and systemic factors on students' opportunities to learn mathematics. Choppin's research focuses on what teachers learn from using innovative curriculum materials, particularly knowledge of how instructional sequences can be used to elicit and refine student reasoning. His current project, the NSF-funded DRK-12 grant Developing Principles for Mathematics Curriculum Design and Use in the Common Core Era (ERGO), focuses on teachers' perceptions and uses of curriculum materials in the context of enacting the Common Core State Standards for Mathematics. He is interested in how digital aspects of curriculum materials transform or reify existing conceptions of curriculum, and how they can be leveraged to increase the emergent and collective development of knowledge in classrooms.

Email: jchoppin@warner.rochester.edu

Jere Confrey is the Joseph D. Moore Distinguished Professor of Mathematics Education at North Carolina State University. She served on the National Validation Committee on the Common Core Standards. She was vice chairman of the Mathematics Sciences Education Board, National Academy of Sciences (1998–2004). She chaired the NRC Committee, which produced *On Evaluating Curricular Effectiveness*, and was a coauthor of NRC's *Scientific Research in Education*. She was a co-founder of the UTEACH program for Secondary Math and Science teacher preparation program at the University of Texas in Austin, and was the founder of SummerMath and co-founder of SummerMath for Teachers. She is the author of numerous pieces of software, led the development of www.turnonccmath.com, a website on the Common Core. Dr. Confrey received a PhD in mathematics education from Cornell University.

Email: jere_confrey@ncsu.edu

Phil Daro served on the writing team of the mathematics Common Core State Standards. He continues to work on implementation and policy issues related to the Common Core. He is the lead designer (mathematics), for the pad-based Common Core System of Courses developed by Pearson Education. He also works in a partnership of the University of California, Stanford and others with the Oakland and San Francisco Unified School Districts for the Strategic Education Research Partnership (SERP), with a focus on mathematics and science learning. Previously, Daro was a Senior Fellow for Mathematics for America's Choice, the executive director of the Public Forum on School Accountability, directed the New Standards Project, and managed research and development for the National Center on Education and the Economy. Daro has directed large-scale teacher professional development programs for the University of California, including the California Mathematics Project and the American Mathematics Project.

Email: phil.daro@gmail.com

Chad Dorsey is President and CEO of the Concord Consortium, which has been an innovation leader in researching and developing STEM educational technology for the past twenty years. Chad's experience ranges across the fields of science, education, and technology. In addition to overseeing a wide variety of STEM projects at the Concord Consortium, he serves as a leader in educational technology across the field on numerous advisory groups and professional workshops. Prior to joining the Concord Consortium in 2008, Chad led teacher professional development workshops as a member of the Maine Mathematics and Science Alliance. Chad has also taught science in classrooms from middle schools through college and has guided educational reform efforts at the district-wide and whole-school levels. While earning his BA in physics at St. Olaf College and his MA in physics at the University of Oregon, Chad conducted experimental fluid mechanics research, built software models of Antarctic ice streams, and dragged a radar sled by hand across South Cascade Glacier. He first met computers when his family hooked an Apple II to their fancy new color TV set, and he's been a shameless geek ever since.

Email: cdorsey@concord.org

A.J. Edson is presently a postdoctoral research associate with the Connected Mathematics Project at Michigan State University. He has been a doctoral fellow in the Center for the Study of Mathematics Curriculum and a research assistant in the Core-Plus Mathematics Project and in the Transition to College Mathematics and Statistics Project. He recently received his PhD in mathematics education in 2014 from Western Michigan University. His dissertation was titled, *A Deeply Digital Instructional Unit on Binomial Distributions and Statistical Inference: A Design Experiment.* His research interests are in secondary school mathematics curriculum design and development, and in efficacy studies focusing on innovative digital instructional materials, with special attention to probability and statistics.

Email: edsona@msu.edu

John Josephus Johnson is the Mathematics Department Chair at Battle High School in Columbia, MO. He has been a classroom teacher for 11 years, serving students ranging from At Risk programs to Honors level while teaching courses from 8th grade through AP Calculus.

Email: johnjohn@columbia.k12.mo.us

Jean-Marie Laborde invented the concept of dynamic geometry in 1985. He studied at the Ecole Normale Supérieure in Mathematics. In 1970, he joined the Centre National de la Recherche Scientifique (CNRS). His doctoral thesis (1977) was devoted to geometric methods for the study of certain classes of graphs, specifically hypercubes, with connection to Automatic Theorem Proving. In 1981 he founded a group of researchers to start

the project Cabri (computerized sketchpad), originally devoted to graph theory. He was appointed in 1994 Director of Research at CNRS leading the EIAH team (Computer Environments for Human Learning). At that time, he developed significant cooperation with Texas Instruments (Dallas) to adapt *Cabri-Geometry SW* for their graphing calculators. He has been a professor and lecturer at various universities in many countries and has supervised more than 15 PhD dissertations. Since 2008, Jean-Marie has led the development of Cabri technology at a new level, offering 2D and 3D direct manipulation in mathematics. He has received various honors including a Doctor Honoris Causa from St. Olaf College (MN) in 2007, and he was Named Knight of the Legion of Honor on Bastille Day, July 14, 2012.
Email: jean-marie.laborde@cabri.com

Brian Lemmen has been teaching high school mathematics for 31 years. He currently is teaching at Holland (MI) Christian High School, a six-year recipient of the Apple Distinguished Program Award. Brian received his BA degree in mathematics in 1983 from Calvin College and his MA from California State University at Fullerton in 1991. He has been involved with the Core-Plus Mathematics Project (CPMP) since 1996 when Holland Christian High School first adopted CPMP. This involvement included being a field-test teacher for the 2nd edition of Core-Plus Mathematics and a professional development facilitator for multi-day workshops for new adopters across the country. More recently, Brian was a field-test teacher for the NSF-funded Transition to College Mathematics and Statistics (TCMS) course. As a field-test teacher he has provided extensive input and feedback to the development of the CPMP-Tools and TCMS-Tools software. Most recently, Brian partnered with A.J. Edson, a PhD candidate at Western Michigan University, in testing a prototype deeply digital unit on Binomial Distributions and Statistical Inference.
Email: blemmen@hollandchristian.org

Hee-chan Lew has been a professor in the Department of Mathematics Education of Korea National University of Education since 1991. He has been a researcher and a research fellow of the Korea Educational Development Institute, the President for the Korea Society of Educational Studies in Mathematics, a member of the International Committee of the IGPME Education, a member of the International Program Committee and a co-chair of the Local Organizing Committee for ICME-12 and now he is a member of the International Program Committee for ICME-13 to be held in 2016 in Hamburg, Germany. He has directed projects in mathematics education on computer technology, teaching methods, evaluation, and textbook development funded by the Korea Research Foundation and Ministry of Education. He is

the author or co-author of Korean elementary and high school mathematics textbooks, 20 research reports and more than 100 articles.
Email: hclew@knue.ac.kr

Valerie L. Mills is the Supervisor and Mathematics Education Consultant for Oakland Schools and current President of the National Council of Supervisors of Mathematics. Oakland Schools is an educational resource center serving 28 school districts and approximately 230,000 students. During her 35+ years in education she has taught high school mathematics, served as Mathematics Department Chair, K–12 Mathematics Coordinator, and Director of Curriculum for the Ypsilanti and Ann Arbor school districts in Michigan. In addition she was the principal investigator on five Mathematics and Science Partnership projects working with high needs districts, was a teacher author on the Core Plus Mathematics Project, president of the Michigan Council of Teachers of Mathematics, past chair of NCTM's Academy Services Committee, and has written numerous professional articles and professional development resources. Mills was awarded the Michigan Mathematics Education Service Award, the Presidential Award for Excellence in Mathematics and Science Teaching, and the Milken National Educator Award.
Email: valerie.mills@oakland.k12.mi.us

David Moursund has a doctorate in mathematics from the University of Wisconsin, Madison. He taught mathematics at Michigan State University and the University of Oregon. He served six years as the first head of the Computer Science Department at the University of Oregon and was a professor in the UO's College of Education for more than 20 years.

His professional career includes founding the International Society for Technology in Education (ISTE) in 1979, serving as ISTE's executive officer for 19 years, and establishing ISTE's flagship publication, *Learning and Leading With Technology*. He was the major professor or co-major professor for 82 doctoral students—six in Mathematics and 76 in Computers in Education. He has authored or coauthored more than 60 academic books and hundreds of articles. In 2007, Moursund founded the non-profit Information Age Education (IAE). IAE provides free online educational materials via its *IAE-pedia*, *IAE Newsletter*, IAE Blog, and books.
Email: moursund@uoregon.edu

Mogens Niss was trained as a pure mathematician at the University of Copenhagen, where he stayed during the first years of his academic career. In 1972 he joined the founding staff of Röskilde University, where he still works. His research interests gradually turned towards mathematics education, especially concerning the justification problem in mathematics education, mathematical applications and modeling, mathematical competencies, assessment, and the nature and development of mathematics education as a research domain. Recently he has become preoccupied with mathematical learning difficulties with high school students. He has been deeply involved in international collaboration in mathematics education, especially as the Secretary General of ICMI (1991–1998), and as a member of the OECD-PISA mathematics expert group (1998–2012). He is currently a member of the Education Committee of the European Mathematical Society. In 2013 he was elected Inaugural Fellow of the American Mathematical Society. He holds an honorary doctorate from the University of Umeå (Sweden).

Email: mn@ruc.dk

Janine Remillard is an associate professor of mathematics education at the University of Pennsylvania's Graduate School of Education. Her research interests include teachers' interactions with mathematics curriculum materials, mathematics teacher learning in urban classrooms, and locally relevant mathematics instruction. She is one of the primary faculty in Penn-GSE's urban teacher education program and is co-editor of the volume, *Mathematics Teachers at Work: Connecting Curriculum Materials and Classroom Instruction*. She is the principal investigator of two NSF-funded studies: Improving Curriculum Use for Better Teaching and Learning About New Demands in Schools: Considering Algebra Policy Environments. Remillard is active in the mathematics education community. She is a research associate of CSMC, chairs the U.S. National Commission on Mathematics Instruction, a commission of the National Academy of Science, and is currently co-chair of SIG-RME.

Email: janiner@gse.upenn.edu

Kenneth Ruthven joined the Faculty of Education at the University of Cambridge after teaching in schools in Scotland and England. He is now Professor of Education and has served as Chair of the Science, Technology, and Mathematics Education group and as Director of Research for the Faculty. His research focuses on curriculum, pedagogy, and assessment, especially in school mathematics, and particularly in respect of the complex and contested process of adaptation to technological innovation. Ken(neth) is former editor-in-chief of *Educational Studies in Mathematics*, recent chair of the British Society for Research Into Learning Mathematics, current chair of Trustees of the School Mathematics Project (SMP), and a fellow of the Academy of Social Sciences (AcSS). Further information, recent projects, and selected publications can be accessed at http://www.educ.cam.ac.uk/people/staff/ruthven/

Email: kr18@cam.ac.uk

Nathalie Sinclair is a full professor in the Faculty of Education, an associate member in the Department of Mathematics and a Canada Research Chair in Tangible Mathematics Learning at Simon Fraser University. She is also an associate editor of *For the Learning of Mathematics* and is a founding editor for a new journal entitled *Digital Experiences in Mathematics Education*. She is the author of *Mathematics and Beauty: Aesthetic Approaches to Teaching Children* (2006) and *Developing Essential Understanding of Geometry for Teaching Mathematics* (2012), among other books. Her primary research concerns the role of digital technologies in the teaching and learning of mathematics, most recently focusing on multi-touch devices and early number sense.
Email: nathsinc@sfu.ca

Kaye Stacey is Emeritus Professor of Mathematics Education at the University of Melbourne, having held the foundation chair there for 20 years. She has worked as a researcher, primary and secondary teacher educator, supervisor of graduate research, and as an adviser to governments. She has written many practically-oriented books and articles for mathematics teachers as well as producing a large set of research articles. Professor Stacey's research interests center on mathematical thinking and problem solving and the mathematics curriculum, particularly the challenges that are faced in adapting to the new technological environment. Her research work is renowned for its high engagement with schools. Her doctoral thesis from the University of Oxford, UK, is in number theory. She was the Chair of the Mathematics Expert Group for the OECD's 2012 PISA survey. Kaye Stacey was awarded a Centenary Medal from the Australian government for outstanding services to mathematical education.
Email: k.stacey@unimelb.edu.au

Michal Yerushalmy is a professor in the department of Mathematics Education at the University of Haifa, Israel. Yerushalmy is the director of the Institute of Research and Development of Alternatives in Education, a member of the Learning in Networked Society (LINKS) National Research Center and Vice President for Research of the University of Haifa. Yerushalmy studies mathematical learning and teaching, focusing on design and implementation of reformed curricula and on cognitive processes involved in learning with multiple external representations, bodily interactions, and modeling. Yerushalmy co-authored and designed numerous software packages in geometry (*The Geometric Supposer*), algebra curriculum (the interactive *VisualMath* secondary school mathematics web curriculum), and studies learning of calculus in dynamic and multi-representation environments. Current projects focus on learning with interactive diagrams in interactive electronic books and on mLearning (mobile learning). In the Math4Mobile project Yerushalmy offers ways to make technology available for mathematical inquiry learning everywhere. Over the past 25 years she has taught courses of didactic methods of mathematics and on cognitive and curricular implications of technology for education. Michal Yerushalmy received the 2010 ISDDE Prize for Excellence in Educational Design.

Email: michalyr@edu.haifa.ac.il

APPENDIX C

CONFERENCE PERSONNEL

PARTICIPANTS

Alcoke, Elizabeth; Francis Parker School; Chicago, IL
Amick, Emily; Think Through Math; Pittsburgh, PA
Baltzey, Patricia; Educational Consultant; Gardiner, MT
Bardeguez, Abner; The University of Chicago; Chicago, IL
Belisle-Chatterjee, Ava; The University of Chicago; Chicago, IL
Benson, John; The University of Chicago; Chicago, IL
Bernstein, Lisa; The University of Chicago; Chicago, IL
Briars, Diane; National Council of Teachers of Mathematics; Reston, VA
Browning, Christine; Western Michigan University; Kalamazoo, MI
Burns, Sarah; The University of Chicago; Chicago, IL
Cagle, Margaret; Vanderbilt University; Nashville, TN
Cappo, Marge; Learning in Motion; Santa Cruz, CA
Carmona, Lisa; McGraw-Hill Education; Columbus, OH
Carter, Andy; The University of Chicago; Chicago, IL
Carter, Kate; University of Chicago Woodlawn Charter School; Chicago, IL
Cengiz-Phillips, Nesrin; University of Michigan–Dearborn; Dearborn, MI
Cirillo, Michelle; University of Delaware; Newark, DE
Clancy, Brian; University of Chicago Woodlawn Charter School; Chicago, IL
Coe, Joanna Sarah; St. Andrew's School; Savannah, GA
Condie, Steven; Illinois Mathematics and Science Academy; Aurora, IL
Cook, Janet; Millard Public Schools; Omaha, NE
Corley, Frank; St. Louis University High; St. Louis, MO

Cox, Dana; Miami University; Oxford, OH
Coyner, Elizabeth; CPM Educational Program; Elk Grove, CA
Dairyko, Ellen, The University of Chicago; Chicago, IL
Darke, Kelly; University of Illinois at Chicago; Chicago, IL
Darrough, Rebecca; University of Missouri – Columbia; Columbia, MO
Davis, Jon; Western Michigan University; Kalamazoo, MI
Dietiker, Leslie; Boston University; Boston, MA
Dingman, Shannon; University of Arkansas; Fayetteville, AR
Drake, Corey; Michigan State University; East Lansing, MI
Economopoulos, Karen; TERC, Cambridge, MA
Engledowl, Chris; University of Missouri–Columbia; Columbia, MO
Fan, Lianghuo; University of Southampton; Hampshire, England
Flores, Kathryn; The University of Chicago; Chicago, IL
Foley, Gregory; Ohio University; Athens, OH
Fonger, Nicole; University of Wisconsin – Madison; Madison, WI
Fonkert, Karen; Charleston Southern University; Charleston, SC
Foster, Eugenie; University of Rochester; Rochester, NY
Frank, Charlotte K.; McGraw-Hill Education; New York, NY
Garfunkel, Sol; COMAP; Bedford, MA
Gauthier, Jason; Allegan ESA and Western Michigan University; Kalamazoo, MI
Gibson, Taylor; North Carolina School of Science and Mathematics; Durham, NC
Goolish, Eric; Adlai E. Stevenson High School; Lincolnshire, IL
Grant, Yvonne; CMP, Michigan State University; East Lansing, MI
Haebig, Caroline; Adlai E. Stevenson High School; Lincolnshire, IL
Hansen, Rob; Scholastic, Inc.; New York, NY
Heid, M. Kathleen; Pennsylvania State University; University Park, PA
Hendry, Neil; International Baccalaureate; The Hague, Netherlands
Hill, Robin; Kentucky Department of Education; Frankfort, KY
Hjalmarson, Margret; National Science Foundation; Arlington, VA
Hocking, Diana; Houghton-Portage Township Schools; Houghton, MI
Hoshi, Chie; Benesse Holdings; Shinjuku-ku, Tokyo, Japan
Hsiao, Joy; Long Island City High School; Long Island City, NY
Huntley, Mary Ann; Cornell University; Ithaca, NY
Jahnke, Anette; University of Gothenburg; Gothenburg, Sweden
Jordan, Laurie; Loyola University; Chicago, IL
Jordan, Steven; Loyola University; Chicago, IL
Kaduk, Catherine; Naperville School District 203; Naperville, IL
Kanter, Patsy; PK Consultants; New Orleans, LA
Kasmer, Lisa; Grand Valley State University; Allendale, MI
Kerzhner, Sofya; Baltimore City Community College; Baltimore, MD
Killian, Courtney; Maternity BVM; Chicago, IL
Kim, Ok-Kyeong; Western Michigan University; Kalamazoo, MI
Kim, Paul; Adlai E. Stevenson High School; Lincolnshire, IL

Conference Personnel ▪ 319

Kissel, Beth; St. Louis University High School; St. Louis, MO,
Koch, Martha; University of Manitoba; Winnipeg, Manitoba, Canada
Lach, Michael; The University of Chicago; Chicago, IL
Lane, Sherry; Quitman School District; Quitman, AR
Larnell, Gregory; University of Illinois at Chicago; Chicago, IL
Leavitt, Della; Chicago Lesson Study Alliance; Chicago, IL
Leimberer, Jennifer; University of Illinois at Chicago; Chicago, IL
Luoma, Jennifer; Houghton High School; Houghton, MI
Machalow, Rowan; University of Pennsylvania; Philadelphia, PA
Mackrell, Kate; University of London; London, England
Males, Lorraine; University of Nebraska – Lincoln; Lincoln, NE
Martin, Catherine; Denver Public Schools; Denver, CO
Martinez, Bertha; Lake Geneva Schools; Lake Geneva, WI
Masters, Robin; Francis Parker School; Chicago, IL
Mateas, Victor; Boston University; Boston, MA
Mathews, Susann; Wright State University; Dayton, OH
Matsunami, Soshi; Meiji University; Kawasaki, Kanagawa, Japan
McConnell, John; ECRA Group; Rosemont, IL
Miller, Christina; US Math Recovery Council; Apple Valley, MN
Millman, Richard; Georgia Institute of Technology; Atlanta, GA
Minor, Beth; McGraw-Hill Education; Richmond, VA
Moran, Cheryl; The University of Chicago; Chicago, IL
Nagaoka, Ryosuke; Meiji University; Kawasaki, Kanagawa, Japan
Neagoy, Monica; Monica Neagoy Consulting Services; Arlington, VA
Nevels, Nevels; Hazelwood School District; Florissant, MO
Nho, Chris; University of Chicago Charter School – Woodlawn; Chicago, IL
Niitsuma, Sho; Meiji University; Kawasaki, Kanagawa, Japan
Okigbo, Carol; Minnesota State University–Moorhead; Moorhead, MN
Olson, Travis; University of Nevada–Las Vegas; Las Vegas, NV
O'Roark, Douglas; The University of Chicago; Chicago, IL
Oslovich, George; Woodstock Community Unit SD 200; Woodstock, IL
Osta, Iman; Lebanese American University; Beirut, Lebanon
Palius, Marjory; Rutgers Graduate School of Education; New Brunswick, NJ
Phalen, Lena; The University of Chicago; Chicago, IL
Pitvorec, Kathleen; The University of Chicago; Chicago, IL
Porter, Denise; The University of Chicago; Chicago, IL
Radhakrishnan, Soundarya; Evanston SD 65, Evanston, IL
Reed, Michelle; Wright State University; Dayton, OH
Reinke, Luke; Amplify; Durham, NC
Revuluri, Sendhil; University of Illinois at Chicago; Chicago, IL
Rich, Kathryn; The University of Chicago; Chicago, IL
Richman, Andrew; Boston University; Boston, MA
Rogalski, Susan; Curriculum Developer; Bedford, MA

Ross, Daniel; Maryville College; Maryville, TN
Rubenstein, Rheta; University of Michigan-Dearborn; Dearborn, MI
Rubin, Jennifer; Houghton High School; Houghton, MI
Ruch, Amanda; The University of Chicago; Chicago, IL
Rudolph, Tammy; Baltimore County Public Schools; Towson, MD
Russell, Susan Jo; TERC; Cambridge, MA
Ryan, Jenna; St. Margaret of Scotland; Chicago, IL
Saputera, Anna; The University of Chicago; Chicago, IL
Sauber, Kristen; Woodstock Community Unit SD 200; Woodstock, IL
Schieffer, Angela; The University of Chicago; Chicago, IL
Scott, Matthew; Millard Public Schools; Omaha, NE
Segal, Stan; Millard Public Schools; Omaha, NE
Senk, Sharon; Michigan State University; East Lansing, MI
Shafer, Kathryn; Ball State University; Muncie, IN
Silver, Edward; University of Michigan; Ann Arbor, MI
Sima, Kathleen; Woodstock Community Unit SD 200; Woodstock, IL
Sizemore, Larry; Baltimore County Public Schools; Towson, MD
Smith, Dana; Woodstock Community Unit SD 200; Woodstock, IL
Smith, David; Utah State Office of Education; Salt Lake City, UT
Spain, Vickie; University of Missouri–Columbia; Columbia, MO
Stadler, Erika; The Swedish National Agency for Education; Stockholm, Sweden
Staley, John; Baltimore County Public Schools; Towson, MD
Stalmack, Richard; Illinois Mathematics and Science Academy; Aurora, IL
Steers, Courtney; Benesse America Inc.; New York, NY
Steingruby, Donald; St. Louis University High; St. Louis, MO
Steketee, Scott; 21st Century Partnership for STEM Education; Philadelphia, PA
Strickland, Carla; The University of Chicago; Chicago, IL
Suurtamm, Christine; University of Ottawa; Ottawa, Ontario, Canada
Switzer, Matt; Texas Christian University; Fort Worth, TX
Takahasi, Akihiko; DePaul University; Chicago, IL
Tarr, James; University of Missouri–Columbia; Columbia, MO
Teuscher, Dawn; Brigham Young University; Provo, UT
Thompson, Denisse; University of South Florida; Tampa, FL
Tollefson, LouAnn; Adlai E. Stevenson High School; Lincolnshire, IL
Tomkiel, Valerie; Adlai E. Stevenson High School; Lincolnshire, IL
Tran, Dung; North Carolina State University; Raleigh, NC
Ulla, Alisha; Spirit of Math Schools; Toronto, Ontario, Canada
Usiskin, Karen; Pearson Education; Glenview, IL
Voytsekhovska, Svitlana; Spirit of Math Schools; Toronto, Ontario, Canada
Wang-Iverson, Patsy; Gabriella & Paul Rosenbaum Foundation; Stockton, NJ
Wargaski, Sarah; Woodstock Community Unit SD 200; Woodstock, IL
Wartowski, David; Niles North High School; Skokie, IL
Whitley, Kevin; Amplify; Durham, NC

Wikstroem, Rose-Marie; University of Gothenburg; Gothenburg, Sweden
Wiland, Kevin; Elgin School District U-46; Elgin, IL
Willert, Lori; Northwood-Kensett Community SD; Northwood, IA
Wiltjer, Mary; Glenbrook South High School; Glenview, IL
Yeong, Joann I; University of Missouri – Columbia; Columbia, MO
Yu, Paul; Grand Valley State University; Allendale, MI
Yuito, Shigeto; Meiji University; Kawasaki, Kanagawa, Japan
Zeller, Erich; MIND Research Institute; Irvine, CA
Ziemke, Kristin; Big Shoulders Fund; Chicago, IL
Zimbler, Idyth; Bernard Zell Anshe Emet Day School; Chicago, IL

SPEAKERS

Asay, Loretta; Clark County School District; Las Vegas, NV
    asayl@interact.ccsd.net

Choppin, Jeffrey; University of Rochester; Rochester, NY
    jchoppin@warner.rochester.edu

Confrey, Jere; North Carolina State University; Raleigh, NC
    jere_confrey@ncsu.edu

Crouse, Angela; McGraw-Hill Education; Chicago, IL
    angela.crouse@mheducation.com

Daro, Philip; Pearson, Berkeley, CA
    phil.daro@gmail.com

Dorsey, Chad; The Concord Consortium; Concord, MA
    cdorsey@concord.org

Edson, Alden; Michigan State University; East Lansing, MI
    edsona@msu.edu

Idol, Chad; McGraw-Hill Education, Chicago, IL
    chad.idol@mheducation.com

Johnson, Josephus; Columbia Public Schools; Columbia, MO
    johnjohn@columbia.k12.mo.us

Laborde, Jean-Marie; Cabrilog—University of Grenoble; Fontaine, France
    jean-marie.laborde@cabri.com

Lemmen, Brian; Holland Christian High School; Holland, MI
    blemmen@hollandchristian.org

Lew, Hee-chan; Korea National University of Education; Cheong-Ju, Chung-Buk, Korea
    hclew@knue.ac.kr

Mills, Valerie; Oakland Schools; Waterford, MI
  valerie.mills@oakland.k12.mi.us

Moursund, David; University of Oregon; Eugene, OR
  moursund@uoregon.edu

Niss, Mogens; Roskilde University; Roskilde, Denmark
  mn@ruc.dk

Remillard, Janine; University of Pennsylvania; Philadelphia, PA
  janiner@gse.upenn.edu

Ruthven, Kenneth; University of Cambridge; Cambridge, Cambridgeshire, UK
  kr18@cam.ac.uk

Sinclair, Nathalie; Simon Fraser University; Burnaby, BC, Canada
  nathsinc@sfu.ca

Stacey, Kaye; University of Melbourne, Melbourne, Victoria, Australia
  k.stacey@unimelb.edu.au

Yerushalmy, Michal; University of Haifa; Haifa, Israel
  michalyr@edu.haifa.ac.il

CSMC LEADERSHIP

Chval, Kathryn; University of Missouri-Columbia; Columbia, MO
  chvalkb@missouri.edu

Hirsch, Christian; Western Michigan University; Kalamazoo, MI
  christian.hirsch@wmich.edu

Phillips, Elizabeth; Michigan State University; East Lansing, MI
  ephillips@math.msu.edu

Reys, Barbara; University of Missouri-Columbia; Columbia, MO
  reysb@missouri.edu

Reys, Bob; University of Missouri-Columbia; Columbia, MO
  reysr@missouri.edu

Smith, Jack; Michigan State University; East Lansing, MI
  jsmith@msu.edu

Usiskin, Zalman; The University of Chicago; Chicago, IL
  z-usiskin@uchicago.edu

Ziebarth, Steven; Western Michigan University; Kalamazoo, MI
  steven.ziebarth@wmich.edu

PROGRAM COMMITTEE

Bates, Meg; The University of Chicago; Chicago, IL
megbates@uchicago.edu

Hirsch, Christian; Western Michigan University; Kalamazoo, MI
christian.hirsch@wmich.edu

Isaacs, Andy; The University of Chicago; Chicago, IL
aisaacs@uchicago.edu

Reys, Barbara; University of Missouri–Columbia; Columbia, MO
reysb@missouri.edu

Shih, Jeffrey; University of Nevada, Las Vegas; Las Vegas, NV
Jeffrey.shih@unlv.edu

Thomas, Amanda; Pennsylvania State University–Harrisburg; Middletown, PA
Alt20@psu.edu

Usiskin, Zalman (chair); The University of Chicago; Chicago, IL
z-usiskin@uchicago.edu

LOCAL ARRANGEMENTS COMMITTEE

Meri Fohran
Martin Gartzman (chair)
Natalie Jakucyn

Lena Phalen
Laurie Thrasher

LOCAL ARRANGEMENTS STAFF

Shaili Datta
Jose Fragoso
Regina Littleton

Caroline Owens
William Schmidt
Michael Stittgen

CPSIA information can be obtained
at www.ICGtesting.com
Printed in the USA
FFOW01n1738280216
21930FF